S0-DTD-972

EXPLORATION

New Lands, New Worlds

DISCOVERING
the
EARTH

EXPLORATION

New Lands, New Worlds

Michael Allaby
Illustrations by Richard Garratt

EXPLORATION: New Lands, New Worlds

Facts On File, Inc.
An imprint of Infobase Publishing
132 West 31st Street
New York NY 10001

Library of Congress Cataloging-in-Publication Data
Allaby, Michael.
Exploration : new lands, new worlds / Michael Allaby ; illustrations by Richard Garratt.
p. cm. — (Discovering the earth)
Includes bibliographical references and index.
ISBN 978-0-8160-6103-7 (alk. paper)
1. Discoveries in geography—Juvenile literature. 2. Explorers—Juvenile literature. I. Garratt, Richard ill. II. Title.

G175.A45 2010
910—dc22 2009031334

Facts On File books are available at special discounts when purchased in bulk quantities for businesses, associations, institutions, or sales promotions. Please call our Special Sales Department in New York at (212) 967-8800 or (800) 322-8755.

You can find Facts On File on the World Wide Web at http://www.factsonfile.com

Text design by Annie O'Donnell
Composition by Hermitage Publishing Services
Illustrations by Richard Garratt
Photo research by Tobi Zausner, Ph.D.
Cover printed by Times Offset (M) Sdn Bhd, Shah Alam, Selangor
Book printed and bound by Times Offset (M) Sdn Bhd, Shah Alam, Selangor
Date printed: June 2010
Printed in Malaysia

10 9 8 7 6 5 4 3 2 1

This book is printed on acid-free paper.

CONTENTS

PREFACE

Almost every day there are new stories about threats to the natural environment or actual damage to it, or about measures that have been taken to protect it. The news is not always bad. Areas of land are set aside for wildlife. New forests are planted. Steps are taken to reduce the pollution of air and water.

Behind all of these news stories are the scientists working to understand more about the natural world and through that understanding to protect it from avoidable harm. The scientists include botanists, zoologists, ecologists, geologists, volcanologists, seismologists, geomorphologists, meteorologists, climatologists, oceanographers, and many more. In their different ways all of them are environmental scientists.

The work of environmental scientists informs policy as well as providing news stories. There are bodies of local, national, and international legislation aimed at protecting the environment and agencies charged with developing and implementing that legislation. Environmental laws and regulations cover every activity that might affect the environment. Consequently every company and every citizen needs to be aware of those rules that affect them.

There are very many books about the environment, environmental protection, and environmental science. Discovering the Earth is different—it is a multivolume set for high school students that tells the stories of how scientists arrived at their present level of understanding. In doing so, this set provides a background, a historical context, to the news reports. Inevitably the stories that the books tell are incomplete. It would be impossible to trace all of the events in the history of each branch of the environmental sciences and recount the lives of all the individual scientists who contributed to them. Instead the books provide a series of snapshots in the form of brief accounts of particular discoveries and of the people who made them. These stories explain the problem that had to be solved, the way it was approached, and, in some cases, the dead ends into which scientists were drawn.

There are seven books in the set that deal with the following topics:

- Earth sciences,
- atmosphere,
- oceans,
- ecology,
- animals,
- plants, and
- exploration.

These topics will be of interest to students of environmental studies, ecology, biology, geography, and geology. Students of the humanities may also enjoy them for the light they shed on the way the scientific aspect of Western culture has developed. The language is not technical, and the text demands no mathematical knowledge. Sidebars are used where necessary to explain a particular concept without interrupting the story. The books are suitable for all high school ages and above, and for people of all ages, students or not, who are interested in how scientists acquired their knowledge of the world about us—how they discovered the Earth.

Research scientists explore the unknown, so their work is like a voyage of discovery, an adventure with an uncertain outcome. The curiosity that drives scientists, the yearning for answers, for explanations of the world about us, is part of what we are. It is what makes us human.

This set will enrich the studies of the high school students for whom the books have been written. The Discovering the Earth series will help science students understand where and when ideas originate in ways that will add depth to their work, and for humanities students it will illuminate certain corners of history and culture they might otherwise overlook. These are worthy objectives, and the books have yet another: They aim to tell entertaining stories about real people and events.

—Michael Allaby
www.michaelallaby.com

ACKNOWLEDGMENTS

All of the diagrams and maps in the Discovering the Earth set were drawn by my colleague and friend Richard Garratt. As always, Richard has transformed my very rough sketches into finished artwork of the highest quality, and I am very grateful to him.

When I first planned these books, I prepared for each of them a "shopping list" of photographs I thought would illustrate them. Those lists were passed to another colleague and friend, Tobi Zausner, who found exactly the pictures I felt the books needed. Her hard work on, enthusiasm for, and understanding of what I was trying to do have enlivened and greatly improved all of the books. Again I am deeply grateful.

Finally, I wish to thank my friends at Facts On File, who have read my text carefully and helped me improve it. I am especially grateful for the patience, good humor, and encouragement of my editor, Frank K. Darmstadt, who unfailingly conceals his exasperation when I am late, laughs at my jokes, and barely flinches when I announce I'm off on vacation. At the very start, Frank agreed that this set of books would be useful. Without him they would not exist at all.

INTRODUCTION

Exploration: New Lands, New Worlds tells of navigators who crossed oceans to chart the coastlines of distant continents, of adventurers who crossed deserts and polar wastes, and of traders who sought new markets and commodities in far lands. As one volume in the Discovering the Earth multivolume set, there is an important sense in which it deals with the topic that underlies all of the others—unraveling the secrets of the planet and its living inhabitants necessitated visiting every part of the world, a task that the navigators and adventurers of old made possible.

The book starts by describing the earliest seagoing ships, the vehicles that would transport diplomats, warriors, and merchants around the Mediterranean region and later around the world. It tells of the Vikings who terrorized Western Europe and colonized Greenland, and of the swift outrigger vessels that sailed from Asia to the islands of the Pacific. Long journeys out of sight of land called for navigational skills, and the book describes the development of navigational instruments such as the sextant and compass, and it explains how to calculate latitude and longitude.

Not all journeys involved ocean crossings. *Exploration* describes the caravans that crossed deserts and the Silk Road network of routes by which goods traveled between Europe and China.

Transporting valuable merchandise by sea attracted predators, seaborne thieves who waylaid vessels. The book explains how they originated and how they operated, and it recounts the lives of a few of the most notorious, including Blackbeard and Captain Kidd.

Many of the great navigators and explorers recorded their experiences. The book describes a few of the most famous, such as John Cabot, Marco Polo, Ferdinand Magellan, Christopher Columbus, James Cook, and Francis Drake. It also tells the story of others who may be less well known, including Pytheas, Xenophon, Friar Odoric, and Ibn Battutah.

Most of the lands the explorers visited possessed resources with a commercial value in Europe. Exploring such lands held out the hope of monetary gain. The Arctic had only one valuable resource—the short route, called the Northwest Passage, between the Atlantic and Pacific. The search for that route stimulated much Arctic exploration. Antarctica had nothing to offer by way of commerce. Its explorers sought only to travel its vast expanses. The book tells of some of the explorers of the world's cold places. It also tells of those who explored the Sahara and the Arabian Desert.

Finally, the story of exploration moves away from Earth and into space. Then it hazards a look into the future. Will people one day live on the Moon and on Mars? Will tourists stay in hotels there? And one day, in the much more distant future, will humans break free from the confines of the solar system and head into interstellar space, on their way to a planet orbiting another star?

Down to the Sea in Ships

People have always been familiar with boats. As recently as the 1950s it was quicker and easier to travel through parts of western Scotland by boat than by road. A glance at a map of Scotland shows that the west is a maze of peninsulas, deep coastal inlets called *sea lochs,* and islands, many of which are inhabited. On land it is impossible to travel any distance in a straight line, because the coast intervenes. Today there are winding roads, augmented by ferry services, and goods that once arrived by sea are now delivered by road, but sailing is still a popular pastime.

The fishing boats that work out of small coastal towns seldom stray out of sight of land. Many of the Scottish ferry routes link places within sight of each other, and the shortest scheduled crossing takes only five minutes. Even the longest, taking several hours, passes between islands, so the sailors remain within sight of land. Exploration involves longer journeys, but for many centuries ships followed coastlines, because they had no means of navigating without landmarks to use as reference points.

Small vessels are adequate for short journeys along rivers or between adjacent ports, but longer journeys call for more substantial ships. They must carry sufficient supplies of food and water to sustain all those on board for the days or even weeks that may elapse between opportunities to replenish stocks. Ships must be large enough to accommodate a crew as well as cargo, and they must be sufficiently robust to ride out bad weather.

This chapter is about the invention of seagoing ships. The story begins in Egypt, where a tradition of building boats to sail the Nile led to the construction of ships that could sail the Red Sea and then the oceans. The chapter describes different styles of ships, including outriggers, Viking ships, galleys, and the biremes and triremes of ancient Greece. It tells of the breakthroughs that came with the invention of the *keel*—the structure that extends from bow to stern along the center of a ship's bottom, adding strength and directional stability to the ship—and rudder. The chapter also describes the voyages made by the Norwegian adventurer Thor Heyerdahl (1914–2002).

EGYPTIANS ON THE NILE

Throughout their long history, Egyptians have depended on the River Nile, and their civilization grew up along its shores. Every year snows melting in the mountains far to the south fed water into the two branches of the river, the White Nile and the Blue Nile, producing a surge that flooded the riverside fields downstream, bringing silt enriched with nutrients to fertilize the crops and water to irrigate them. In addition, Egyptians ate fish that they caught from boats on the Nile, and from the very earliest times the river was the thoroughfare that linked communities.

Timber was a scarce commodity in ancient Egypt and most of it had to be imported, but *papyrus* was abundant. Papyrus *(Cyperus papyrus)* is a *sedge*—a flowering plant (family Cyperaceae) resembling grass and rush—that grew in the wetlands of the Nile Delta. Papyrus plants are up to nine feet (2.7 m) tall and bundles of them, tied tightly together, are waterproof. Egyptians and other Middle Eastern peoples used papyrus to make mats, mattresses, paper—and boats. Papyrus boats were made from bundles of papyrus tied together with rope. At first they were simple rafts that were quick to make, but later people made papyrus boats with raised sides and a high stem and stern. Some boats had masts and sails, and even deckhouses. They could be powered by sail or rowed; some were towed from the riverbank, and others were allowed to drift with the current. According to the Greek historian Herodotus (ca. 484–ca. 425 B.C.E.; see "Herodotus and his Travels" on pages 107–109), the boats that drifted had a crate, shaped like a door and made from wood and reed mats, that floated ahead of the boat attached by a rope, and a stone with a hole drilled through it attached by another rope to the stern. The current swept the crate along, pulling the boat behind it, while the stone dragging in the rear held the boat on a straight course. Some of these "drifters" were able to carry heavy loads. Large stone sculptures traveled the Nile on riverboats.

The larger of these vessels were seaworthy, at least in fine weather, and could venture beyond the river (see *Kon-Tiki, Ra,* and *Tigris* on pages 25–32). Later ships, capable of longer sea voyages, were made from timber. In about 1490 B.C.E. Queen Hatshepsut (reigned in her own right 1473–1458 B.C.E.) sent a trading expedition to the Land of Punt, in the Horn of Africa, with additional instructions to collect animals and plants. The expedition consisted of five ships, each of them 70 feet (21 m) long and 16 feet (5 m) wide, with a sail and 30 rowers. The illustration below shows an Egyptian merchant ship from about 1250 B.C.E., but this tried-and-true design remained in use for a long time, and the ship in the picture was the same size as the ones

The type of ship used by the Egyptians on the Red Sea in about 1250 B.C.E. The thick rope between the bow and stern is under tension and helped make the ship stable. The single mast held a sail 50 feet (15 m) wide. There were 15 rowers on each side, and two oars fastened together at the stern acted as a rudder. The carving on the stern is of a lotus flower.

that sailed to Punt 240 years earlier. The sail was rectangular, as were all Egyptian sails. It was 50 feet (15 m) wide and was held between two spars. There were 15 rowers on each side and two oars lashed together at the stern served as a rudder. The ship had no keel to give it structural strength; instead there was a thick rope running from bow to stern between the two ranks of rowers. This rope was held under tension by twisting a strong pole inserted through its strands. Some ships used a raised wooden gangway instead of the rope.

Ships of this design were unable to sail into the wind (see "The Keel, and Sailing Into the Wind" on pages 17–18), and had to be rowed for much of the time. Consequently, they required a large crew. Their reliance on oars may make them appear archaic, but European *galleys*—seagoing ships that could be rowed—were still in use in the late 17th century.

The Egyptians also built much larger ships, suitable for longer sea voyages, and they had warships. These had raised sides to protect the rowers, sailors, and soldiers, and nine oars on each side. By about 600 B.C.E. the Egyptians were building large warships, capable of ramming enemy vessels, with rowers on two or more levels.

OUTRIGGERS

When, in 1521, the Portuguese explorer Ferdinand Magellan (1480–1521; see "Ferdinand Magellan, from Atlantic to Pacific" on pages 132–135) reached the Mariana Islands in the Pacific Ocean, the islanders came out to meet him in sail-powered boats that were faster and more maneuverable than the ships he commanded—and some of them were longer. Historians believe that the islanders reached the Marianas in about 2000 B.C.E., arriving in vessels very like those that greeted Magellan. The other Polynesian peoples who sailed from Southeast Asia more than 1,000 years ago, eventually to colonize all the habitable islands of the South Pacific, also traveled in boats made to a similar design: the *outrigger canoe.* This is a small, narrow boat that is stabilized by one or two long floats, the outrigger(s), fastened by rigid struts to the main hull. Traditionally, the main hull on the smaller vessels was a *dugout canoe,* made by hollowing out a straight tree trunk. In larger outrigger canoes the main hull was made from planks. Canoes with a single outrigger were the more common type, and those with two outriggers were not used for long ocean voyages.

Outrigger canoes sail with the single outrigger on the *windward side*—the side exposed to the wind—and the main hull on the *lee side*—the side sheltered from the wind. The outrigger's weight prevents the craft from overturning, and its location on the windward side of the boat helps maintain directional stability.

The double canoe was an alternative to the outrigger canoe. This comprised two identical canoes connected by struts, usually with 12–30 inches (30–75 cm) between the two boats. The connecting struts were the most important component, for each canoe was too narrow to be stable by itself. Should the struts fail, both canoes were doomed.

In 1774 the small fleet commanded by James Cook (1728–79; see "James Cook and Scientific Exploration" on pages 142–145) reached Tahiti. Johann Reinhold Forster (1729–98), the expedition's official naturalist, recorded seeing 159 double canoes, each one 50–90 feet (15–27 m) long, and 70 smaller double canoes lying offshore. These were war canoes, with platforms for warriors, and the smaller canoes had a roof or cabin at the stern. Forster recorded that even the smallest district of Tahiti possessed 40 of the larger vessels. The Dutch explorer Abel Tasman (1603–59) reported seeing only double canoes during his voyages around New Zealand. Cook saw double canoes along the coasts of South Island, but only one off North Island. In fact, double canoes were used throughout the Pacific at that time, and some were much bigger than those Forster saw at Tahiti. Sailors from Samoa and the Cook Islands had double canoes that were up to 150 feet (45 m) long.

Four years later, on January 20, 1778, the Cook expedition arrived at Kauai Island, Hawaii. Prior to this, other Europeans had been shipwrecked or marooned on the Hawaiian Islands, but they had all either died or settled there. Cook's party was the first to reach the Hawaiian Islands and return home from them. Cook named these islands the Sandwich Islands, to honor his patron, the earl of Sandwich. As his ship, HMS *Resolution,* entered the bay, it was greeted by more than 3,000 outrigger canoes, finished to a standard the English carpenters and cabinetmakers admired, being paddled by more than 15,000 men, women, and children. What Cook could not know was that a Hawaiian tradition held that long ago the god Lono had taken human form and departed, but one day he would return. The Hawaiians thought that the *Resolution* was Lono's boat, Cook was the incarnation of Lono, and they were witnessing the god's return.

James Cook estimated that an outrigger canoe could attain a speed of 22 knots (25 MPH, 40 km/h) under favorable conditions and could cover 120 or more miles (193 km) in a day. Vessels built for long voyages could remain at sea for many days. The largest outrigger canoes could carry up to 50 people and a 60-foot (18-m) canoe could carry three tons (2.7 tonnes) of cargo.

Outrigger canoes are still widely used and today racing them is a popular sport. In Sri Lanka they are used for commercial sea fishing. A few have engines, but these are costly, and most Sri Lankan outrigger fishing boats are nonmotorized. Many are constructed in the traditional way, with a dugout main hull.

THE FLYING PROA

The outrigger canoe reached the pinnacle of sophistication with a version called the proa. The word *proa,* or something very like it, means "boat" in most of the languages spoken in Polynesia and Micronesia. It was the fastest sailing vessel ever built, and it achieved its remarkable performance by employing a unique design. When a proa called the *Amaryllis* appeared at an American regatta in 1876, the *New York Times* published (June 26) the following description:

> The fiercest squall cannot capsize a flying-proa, even if she is handled by a Presbyterian minister from an inland town. . . . If her two hulls are made of galvanized iron divided into watertight compartments, she might strike on every rock in Hell Gate without sustaining any fatal injury; and while her light draught would render her fast before the wind, the inner side of the weather hull, when on the wind, would have a greater hold on the water than has the ordinary centre-board. . . . The success of the *Amaryllis* shows that as a racing machine she is as much superior in model to the fastest keel or centre-board boat, as the latter is to a mud-scow. Her extraordinary speed, however, is not her best quality. . . . To sail a vessel like the *Amaryllis* requires about as much seamanship as is needed to handle a wheel-barrow.

The hull of a traditional proa comprised a dugout canoe with the sides built up by planking, usually by about five feet (1.5 m). The hull's unique feature was its asymmetry. One side was curved, bulging like

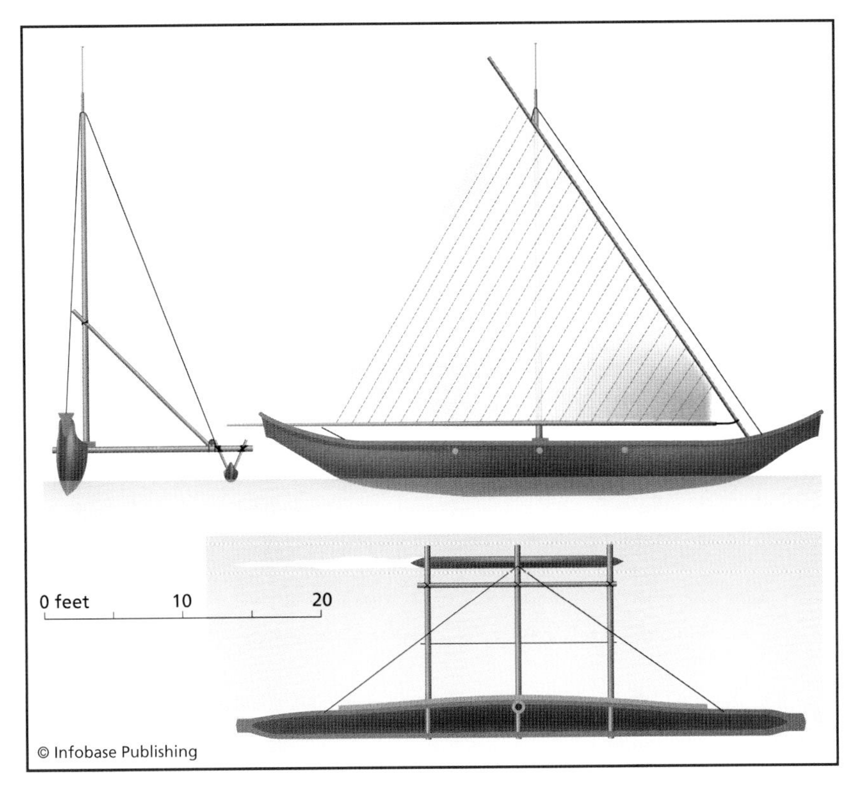

Until the 20th century, the proa was the fastest sailboat the world had ever known. One side of the hull is rounded like that of an ordinary canoe and the other side is completely flat. The proa sails with the rounded side always to windward. An outrigger on the windward side provides stability. The bow and stern are the same shape. The drawings show a proa head-on *(left)*, from the side *(middle)*, and in plan view *(right)*; the scale bar is 20 feet (6.1 m) long.

the side of an ordinary canoe, but the opposite side was completely flat. The vessel was always sailed with the curved side facing into the wind, so the proa always sailed at right angles to the wind. The bow and stern were identical in shape. The outrigger, called the *ama* in most Polynesian languages, was made from a hollowed log shaped like a small boat and it was attached to the windward (curved) side of the hull by a frame made of bamboo poles. The mast was positioned halfway between the bow and stern, but on the central strut of the frame to the outrigger, so it was on one side of the boat. The large, triangular sail was attached to a yard, the lower end of which fitted into a socket close to the bow, and the bottom edge of the sail was attached to a boom. Both the yard and boom were bamboo poles. This arrangement held the sail almost flat. The amount of sail could be adjusted according to the wind by rolling it around the boom. The illustration om page 7 shows a traditional proa with its sail set, seen head-on, from the side, and in plan view.

The proa always sailed with the outrigger on the windward side, so when it was necessary to reverse direction the crew would turn the proa until its stern was into the wind, then raise the yard from its socket, carry it to the opposite end of the boat, and place it into the identical socket at that end. The bow and stern had then reversed positions and with appropriate adjustments to the sail, the proa was ready to sail in the return direction.

The *Times* reporter may have underestimated the skill required to sail a proa at full speed in a strong wind. As the wind pushed the boat from the side, the crew would balance it by moving onto the frame between the hull and outrigger. Their aim was to allow the main hull to lean over far enough for the outrigger to leave the water so it skimmed along the surface, greatly reducing drag. Because the outrigger was out of the water, this was called flying, and it is why the vessel was called the flying proa. Proas were made in a range of sizes. Many were about 15 feet (4.5 m) long, but there were others up to 100 feet (30 m) in length and much smaller ones that children could manage.

ROMAN GALLEYS

Ancient Rome controlled a large empire. The authorities needed to move troops and officials over long distances and Roman merchants traded with all the subject territories. The Mediterranean Basin lay at the heart of the empire, and so a great deal of Roman traffic traveled by sea. Inland, large, navigable rivers, such as the Tiber, Danube, Rhine, and Nile, penetrated deep into Roman territories, and Roman military and merchant ships sailed on them. The Roman army was respected and feared everywhere, but Rome also had a formidable navy.

Military seagoing ships were galleys—ships propelled by sail when the wind was favorable and at other times by very long oars. The oars were necessary because Roman ships carried rectangular sails and had a shallow draft, which meant that they were unable to tack into

Roman ships used one or two rectangular sails and they could not sail into a headwind. Consequently, they used oars manned by 50–80 oarsmen. There were large steering oars at the stern (at left in this picture), a castle from where the captain had a clear view of the entire ship, and a powerful ram at the bow. This type of ship was called a liburnia.

a headwind. Although the sail could swing on its yard, if it turned to catch a wind from the side the pressure would capsize the ship.

The illustration above shows a Roman warship of a very successful type called a *liburnia*. The liburnia had a single sail—some warships carried two—and the one shown here had two tiers of 11 oars on each side, making 44 oars in all. At its bow there was a strong ram. Prior to the invention of naval guns, warships fought by ramming, hoping to hole the enemy vessel below the waterline. Metal plates at the bow and stern provided some protection against ramming. The ship was steered by a large oar at the stern. From the top of the castle the captain had a clear view of the entire vessel. In addition to its oarsmen—several to each oar—a liburnia carried up to 50 soldiers. A liburnia had a deck, which allowed it to carry more soldiers than would have been possible on an open ship. If ramming failed, the attacking ship would try to pull alongside its opponent so its soldiers could swing out a gangplank, allowing them to board the enemy ship and engage its crew in hand-to-hand combat. Ships of this type were seaworthy and they also sailed the major rivers.

The Roman navy also used *quinqueremes*—vessels with five ranks of rowers. These ships were armed with catapults capable of hurling firebombs and they could also carry up to 120 soldiers. Quinqueremes were formidable weapons in the Roman arsenal.

A different design was used for the ships that carried military provisions. This type of ship had very high sides and a three-pronged or trident ram. Its interesting feature was that its oars were arranged in three groups of four on each side, with each group on a balcony, called a *crinoline,* at a different height. Each group of oars could be used independently of the others and there were two large steering oars near the stern, one on each side of the hull. This construction made the ship highly maneuverable in small spaces. At sea, when the wind was from the stern, the ship carried a square sail on a mast at the stern. The mast was removed when the sail was not in use.

Merchant ships were wider than warships and they did not use oars—they were not galleys. They had a rounded, very robust hull with a mast at the center carrying a rectangular sail. Two triangular sails fitted on either side of the mast above the yard carrying the main sail increased the sail area. A second inclined mast near the prow and projecting forward carried a small, square sail. This sail could turn to catch a wind from the side, improving the ship's performance. Two large steering oars were mounted on crinolines at the stern.

Some merchant ships were very big. Rome imported grain and its grain ships carried up to 1,000 tons (900 tonnes). In 1907 a sponge diver discovered the wreck of a Roman ship at Mahdia, Tunisia. It was about 130 feet (40 m) long and still had its cargo of 70 marble pillars. The largest Roman merchant ship known, nicknamed Caligula's giant ship, was discovered in the 1950s during the construction of Rome's

Fiumicino airport on the site of the ancient port of Ostia. That vessel was 312 feet (95 m) long, 69 feet (21 m) wide, and could probably have carried a cargo weighing 1,300 tons (1,180 tonnes).

BIREMES AND TRIREMES

Greece is a land of islands and from earliest times its city-states fought frequent and bitter wars. The earliest Greek warships varied in size and shape, but from about 800 B.C.E. they began to be built with rams and from that time warship design diverged from the design used for merchant vessels. The first warships had only one rank of oars, usually about 25 on each side. The largest of them were up to about 120 feet (37 m) long and about 13 feet (4 m) wide.

An improvement came when ships carried two ranks of oars, one above the other and later designs added more ranks. A galley with two rows of oars on each side was called a *bireme;* one with three rows of oars was a *trireme;* one with four rows was a *quadrireme;* and one with five rows—favored by the Romans—was a quinquereme.

A bireme was narrow—about 10 feet (3 m) wide—and fast. Its ram was in the shape of a trident or of a wild boar's head and above the ram there was a strong wooden block with a hole through which a rope could be passed to tie several ships together, allowing them to attack in a close formation. The vessel did not have a closed deck, but a crinoline ran along the center, supported on beams from side to side of the ship. The upper rank of oars passed over the ship's *gunwale* (pronounced "gunnel")—the upper edge of the ship's side—and the lower rank passed through holes in the planking sealed by leather to prevent water entering.

The basic bireme design remained in use for at least 800 years, but it had a major disadvantage: The two ranks of oars often caught each other and became interlocked, and untangling them was a long and difficult task. Versions introduced from about 700 B.C.E. solved the problem by fitting outriggers extending on both sides of the ship, but not touching the water. The upper rank of rowers sat on benches along the outriggers, with their oars well clear of those of the lower rank. There was one oarsman to each oar. Most biremes had about 26 oars on each side, but large ships had many more. In addition to the oarsmen, a bireme carried a small number of archers and foot soldiers. If the enemy boarded the ship the soldiers would aim to hold them at bay while the oarsmen seized weapons and shields and came to join them.

In about 650 B.C.E. the first triremes were built in Corinth, and in little more than a century all of the city-states were using them. The hull was still narrow and the outriggers carried the upper two ranks of oars, the lowest oars passing through the hull. A trireme was approximately 115 feet (35 m) long, but because of the outriggers the hull was only about 11 feet (3.5 m) wide. It carried 170 rowers, and with a crew in peak condition a trireme could maintain an average speed of nine knots (10.3 MPH, 16.6 km/h) for 24 hours. It was a heavy ship, used for ship-to-ship combat with the ram. On January 1, 1987, the Trireme Trust, based in Britain, launched a replica of the Athenian trireme *Olympias,* the original of which was built in about 400 B.C.E. The *Olympias* is 120 feet (37 m) long and manned by volunteers. The illustration below shows the ship just after its launch near the island of Poros.

Although slaves were sometimes employed as rowers, this happened only when no free men were available. It was an honor to be an oarsman on a Greek warship. Rowers had to be physically fit and

A life-size replica of an Athenian trireme photographed on January 1, 1987, off the island of Poros. British volunteers are manning the oars. *(Susan Muhlhauser/Time & Life Pictures)*

strong—in tests on reconstructed triremes, modern rowers could not match the performance of ancient Greek oarsmen as reported by historians. They were also highly trained. It was vital that the movements of the oars were synchronized to maximize efficiency and to prevent oars clashing and becoming entangled.

During the Hellenistic period (323–146 B.C.E.), when the Greeks were at the height of their influence, the largest warships were *catamarans*—vessels with two hulls—manned by up to 4,000 oarsmen and equipped with many catapults and a large ram. These were powerful, but extremely expensive to build, and they were slow. Smaller, faster, more maneuverable ships were able to avoid them, and when the smaller ships attacked in packs, there was a high probability that at least one or two of them would survive the catapult bombs for long enough to ram the catamaran and sink it.

MERCHANTMEN AND WARSHIPS

Warships must be fast, maneuverable, and well armed. Alternatively, they may sacrifice speed and maneuverability in order to carry much heavier armament. In ancient times, however, as now, neither of these specifications was appropriate for most of the ships that were plying the seas. Despite the warlike natures of rival city-states and empires, most of the time the need was not for warships but for cargo ships working among the islands and hugging the coast on short voyages, occasionally venturing out of sight of land. In the time of the Roman emperor Diocletian (244–311 C.E.) it was cheaper to transport grain by sea from one end of the Mediterranean to the other than to move it 75 miles (120 km) overland by cart. Moreover, once Rome had consolidated its military dominion over the Mediterranean and the navigable rivers, no rival empire was capable of challenging the might of the imperial navy, so warship design became less important than it had been earlier.

Merchant ships were rounder in cross section than the sleek warships were. Most Greek and Roman merchantmen were powered entirely by sail. Roman cargo ships had a main mast in the center, carrying a large, rectangular sail. The *prow* was low. The prow is the part of the ship that projects forward from the *stem*—the curved piece of timber at the forward end of the ship that is an extension of the keel and to which the sides of the ship are attached. The *stern*—

the rearmost part of the ship—was high, to allow the steersman a clear view of what lay ahead of the ship.

Their reliance on sail meant they were slow and often becalmed. If such a ship sailed against the prevailing wind it might maintain an average speed of no more than two knots (2.3 MPH, 3.7 km/h), although with fair winds it could maintain up to nine knots (10 MPH, 17 km/h). On the other hand, a sailing ship did not have to accommodate a large crew of oarsmen. These ships carried timber, which was being bought and sold around the ports of the eastern Mediterranean and Egypt by 1000 B.C.E., grain, and other imperishable goods including metals, statues, and ornamental stone (building stone was always quarried close to where it was to be used). Athenian merchant ships could probably carry up to about 160 tons (145 tonnes) of grain, but some Roman ships could carry more than double that. In the center of St. Peter's Square in Vatican City stands a granite obelisk, 84 feet (25.5 m) tall, which originally came from Egypt. It was transported to Rome in 37 C.E. on the order of the emperor Caligula (12–41 C.E.). The obelisk weighs approximately 1,456 tons (1,323 tonnes), so that must have been the capacity of the ship that carried it.

There were exceptions, however, and some cargo ships were galleys. These vessels resembled warships in design, but they were more strongly built and they carried no soldiers, rams or catapults. Their greater speed meant they could carry passengers and perishable goods such as wine and olive oil—high-value cargoes that would have offset the cost of the larger crew. Galleys were also better at negotiating crowded ports and at traveling along rivers. In order to serve river ports, ships relying only on sails had to transfer their cargoes to riverboats, which introduced further delays and added to the cost.

MASTS AND SAILS

Everyone who goes outdoors on a windy day knows that the wind exerts pressure, so it is possible that thousands of years ago sailors learned that if they stood up while holding out their garments, the wind would help propel their boats. It is possible, but no one can know for sure, because there is no record. In any case, it would not have taken long for the sailors to tire of standing in this way and to figure that an expanse of fabric attached to a pole—a sail attached to a mast—would work better. There are records of early Egyptian boats (see "Egyptians on the Nile" on pages 2–4). These were made from bundles of reeds and some of them had masts in the shape of an upside-down V, which was the shape best suited for a reed boat. Other Egyptian vessels had a single mast, and from about 2200 B.C.E. the single mast was the only type used.

The mast was *stepped*—fitted into a socket attached to the hull. It was positioned toward the bow and held in place by ropes. When the sail was not in use it was detached from the mast and stowed away. The mast was then unstepped and laid on the deck with its top resting on a support at the stern. Over the centuries the position of the mast gradually moved toward the stern until, by about 1500 B.C.E., it was approximately in the center of the boat.

Egyptian sails were rectangular and very large. At first they were taller than they were wide, but in later designs the sails were wider than they were tall. A ship had a single sail that hung from a *yard*—the rigid support along the top. It was held in shape by a *boom*—a rigid support along the bottom edge of the sail. Sailors on the deck raised and lowered the sail by means of ropes called *halyards*. The earliest sails were made from woven papyrus, but in time papyrus was replaced by more durable linen.

The prevailing winds over Egypt blow from the north or northwest. These are ideal for ships sailing upstream along the Nile, and square or rectangular sails are the best shape for use in a wind from behind. Egyptian riverboats used their sails to travel upstream and returned with their masts and sails stowed, drifting with the current or being towed from the riverbank.

Square sails are also efficient on the open sea. Although their yards are able to turn against the mast, they are pivoted at the center so they never sweep across the ship and the wind always pushes against the same side of the sail. This makes square sails safer to use than *fore-and-aft sails*—sails that can take the wind on either side by swinging across the ship.

The disadvantage of a square sail is that a square-rigged ship is much slower than a ship with fore-and-aft sails when sailing into a headwind. A wind that blows exactly in the same or opposite direction to that of the ship exerts the whole of its force either for or against the ship's motion. If the ship's sails can be turned at an angle to a headwind, however, a component of the wind force acts in the direction the ship is following. Sailors use this fact to *tack*—change direction in order to allow a headwind to blow from the side so the ship can advance against it. By repeatedly tacking, so that wind blows alternately from the *port* (the left side facing forward) and *starboard* (the right side facing forward) the ship advances against the wind along a zigzag path, as shown in the illustration below. A square sail has restricted movement, however, which means that a square-rigged ship cannot sail at an angle of less than about 70° to the wind direction. A ship with a fore-and-aft rig, in contrast, can sail at about 40° to the wind direction. Consequently, when tacking, a square-rigged ship follows a much tighter, and therefore longer, zigzag path than a fore-and-aft-rigged ship. This has never been especially important on long voyages, because the commander or *sailing master*—the member of the crew responsible for navigation and determining the deployment of sails—would plan a route where the prevailing winds would be from behind.

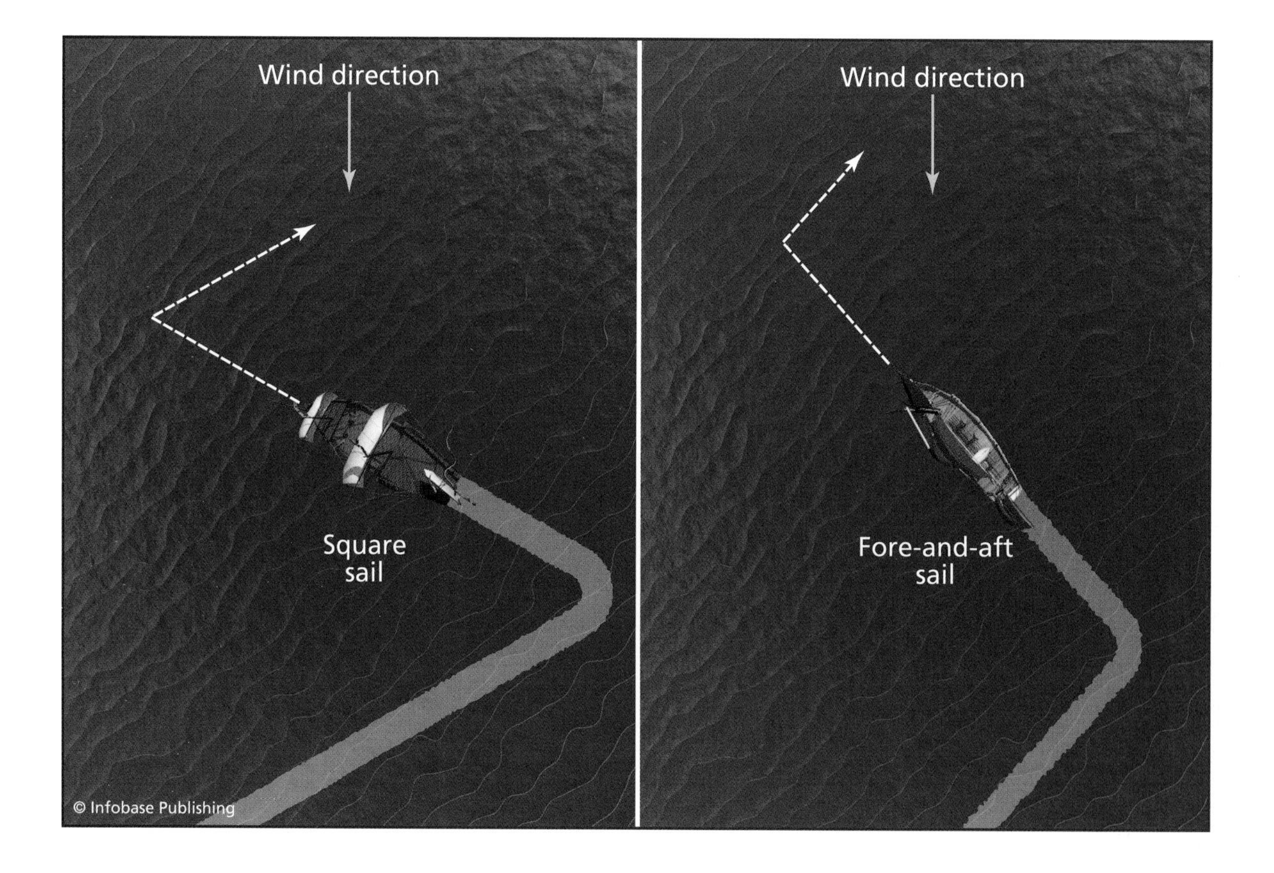

A ship with a square sail cannot travel at an angle less than about 70° to a headwind. A ship with a fore-and-aft sail can travel at an angle of about 40°.

The lateen rig first appeared in the eastern Mediterranean in the second century C.E. Chinese junk-rigged ships were sailing the oceans at about the same time, but were invented during the Han dynasty (200 B.C.E.–200 C.E.). The gaff rig was invented in Europe in the 17th century.

The illustration above shows three of the basic fore-and-aft rigs. The earliest was probably the *lateen rig*—a triangular sail held on a long yard with the forward end very low and the aft end high, held on a mast that is often inclined forward. This type of rig first appeared in the eastern Mediterranean in the first century C.E., and the name may be a corruption of *latin.* It proved very practical and efficient. Christopher Columbus (1451–1506) commanded a three-masted *caravel,* which was a two- or three-masted ship with lateen sails on all masts (see "The Voyages of Christopher Columbus" on pages 147–152). The gaff rig was invented in Europe in the 17th century.

THE KEEL, AND SAILING INTO THE WIND

The bigger a ship is, the more it can carry, but in ancient times this presented a problem. When a ship exceeded a certain length that varied according to the materials used to build it, it had a tendency to sag in the middle, and sagging was often a prelude to the vessel breaking apart. Thus, ships could not be made too long, and that limited the quantity of goods or number of passengers they could carry. The Egyptians dealt with the problem by passing a strong rope around the ship, from stem to stern, to hold it together under tension (see "Egyptians on the Nile" on pages 2–4). Then someone had a better idea.

Early in the 14th century B.C.E. a ship sank off Ulu Burun on the southern coast of what is now Turkey. In 1982 a sponge diver called Mehmet Cakir discovered the wreck. The Uluburun ship (as it is now called) was about 50 feet (15 m) long and had been carrying a cargo of tin, glass, ebony, ivory, amber, gold, lamps, tools, weapons, and many other items. Its cargo also included approximately 11 tons (10.3 tonnes) of copper ingots. The copper and tin were obviously intended for making bronze, but most copper compounds are toxic and those from the Uluburun wreck killed any organism seeking to feed on the ship's timbers. The copper preserved parts of the ship, allowing archaeologists to study its construction.

The Uluburun ship was built on a keel that was 11 inches (27.5 cm) wide. The hull was made from planks about 2.4 inches (6 cm) thick joined to the keel and to each other by mortise-and-tenon joints secured by wooden pegs, about 0.9 inch (2.2 cm) in diameter, driven through the tenons. It was built from cedar (*Cedrus* species) timber. This is the oldest ship that is known to have possessed a keel, but knowledge of the technique gradually spread.

By about 1250 B.C.E. contemporary paintings show that the Sea Peoples were sailing ships with keels. (The Sea Peoples were seafaring raiders who roamed the eastern Mediterranean and attempted to seize Egyptian territory. Historians are uncertain where they originated.)

A keel is the first part of a ship to be made. The hull is then built onto the keel on either side. This provides the structural strength needed to prevent the hull from sagging when loaded. The keel also serves two additional functions. By its nature, it is large and heavy. It adds weight to the ship's bottom, lowering the center of gravity and making the ship less likely to capsize in a sidewind. It also projects below the vessel along the entire length of the hull, helping to prevent the ship from drifting sideways. It was the invention of the keel that allowed ships to sail against the wind. If a ship without a keel tried to tack into a headwind, the wind would simply drive it back, drifting at an angle.

PORT, STARBOARD, AND THE INVENTION OF THE RUDDER

The introduction of the keel brought a further advance. At either end of the ship the keel is joined to posts at the stem and stern. These posts rise to the top of the hull and in ancient ship designs they often continued higher, ending as carved symbols or ornaments. The stem post strengthened the ship's bow and in later designs it provided support for the *bowsprit*—a pole that extended forward from the prow of a sailing ship and to which the stays from the foremast could be attached, allowing the foremast to be positioned farther forward than would be possible otherwise.

The stern post provided a strong, rigid structure that could hold a rudder. On earlier ships the steersman used an oar to hold the vessel on a constant heading and to change its direction. At first the steering oar was an ordinary oar, identical to those used to propel the ship. It was located on one or other side of the ship, and during the Old Kingdom period in Egypt (2686–2134 B.C.E.) some ships had as many as five steering oars on each side as well as one attached to the stern. More usually, steering oars were fixed to one or both *quarters*—the *quarter* is the part of the ship approximately one-fifth of the distance from the stern to the bow. A steersman operated the oar by means of a horizontal pole inserted through the stock of the oar to provide leverage. As ship design advanced, it became possible to reduce the number of steering oars, eventually to one. Steering oars remained in use until the Middle Ages. There are carvings dated at about 1180 C.E. in churches in Belgium and England that depict ships with rudders hinged to their stern posts. These are the first dated references to rudders.

Steering oars were effective at controlling ships, but they had significant disadvantages. The steersman working an oar had to be able to move freely over a fairly large area of the ship's deck. At times this could interfere with the sailors adjusting the sails by means of ropes. Steering oars located on the side of the hull also caused drag, slowing the ship. The rudder was a great improvement. It provided better directional control while minimizing drag. It was also easier to use, because it could be attached to a *tiller*—a pole inserted horizontally through the rudder shaft—or to two ropes, and, much later, by chains to a helm wheel. Ships were certainly equipped with helm wheels by about 1700; the wheels were probably introduced in the second half of the 17th century.

In Anglo-Saxon the steering oar was known as the *steorbord,* and throughout Northern Europe it was always attached to the right-hand side of the ship as seen looking toward the bow. In time *steorbord* became the English word *starboard,* and *starboard* remained in use after the steering oar had given way to the stern-mounted rudder. It was important not to damage the steering oar, so when a ship tied up at a quayside it always did so with the left side of the ship next to the quay. That side of the ship was known as the *bœbord* in Old English, perhaps because the steersman had his back to that side. The word has survived in German as *Backbord* and in French as *bâbord,* but it did not survive in later versions of English. The side adjacent to the quay was the one from which the ship would be loaded, or *laded* in Middle English, so this side became the *laddeborde* or *ladeborde,* which later developed into *larboard.* The trouble was that when commands were shouted above the noise and bustle of a working sailing ship, *larboard* could sound just like *starboard.* In the middle of the 19th century, therefore, the British Admiralty issued an Admiralty order requiring that henceforth the larboard side of the ship should be known as the *port* side. When airplanes came into military and commercial (rather than purely private) use, aviators adopted the same terms, so the sides of any airplane are known as starboard and port, as are the wings and engines.

LONGSHIPS

From the eighth to the 11th centuries the Vikings of Scandinavia terrorized coastal communities throughout western Europe. The sight of the approaching longships, with their sleek lines, rectangular sails, high dragon-headed prows, and waving pennants would send villagers running from their homes to hide in the hills or forests. The longships were not warships, but troop transports. They brought warriors seeking plunder. Their victims called the longships "dragon ships." The illustration on page 20 shows what the approach of an invading Viking fleet might have looked like as the longships sped into an English estuary. Their shallow draft allowed these vessels to sail far upstream, and if there were settlements inland from the coast or some distance from a river, the warriors would raid farms for horses

An artist's impression of a fleet of Viking longships speeding up an English estuary in the 10th century *(Hulton Archive/Getty Images)*

and head on horseback for the nearest village or church. Churches were often targets not on religious grounds but because the Vikings knew they usually contained valuable items.

There were several types of longship, but all of them were built in the same way. Their design had evolved over many centuries from that of a dugout canoe. They were long, narrow, and shallow—the shape of a tree trunk. The shape of a ship is often given as the ratio of its length to its width. Longships were never less than six times longer than they were wide (6:1). In 1935 archaeologists examining the remains of a ship in which a chief had been buried, found the ship had been 68 feet (20.6 m) long, 10 feet (3.2 m) wide (ratio 6.8:1), and 3 feet (1 m) high, measured vertically from the keel to the gunwale. The timbers of a longship discovered in the harbor at Hedeby, on the Danish–German border, in 1953 showed that the vessel had had a ratio of 11.4:1. Archaeologists determined that the Hedeby ship had been a fire ship that had been set alight and sent blazing toward the town during an attack in about the year 1000.

Longship construction technique reached its peak in the ninth century. Work began with making the keel, which was T-shaped in cross section, and the posts at the stem and stern. The vessels were *clinker built*—made from planks the length of the vessel, with each plank overlapping the one below it. Starting at the bottom, the *strakes*—planks that were joined end-to-end to build the hull—were fastened first to the keel and then each layer was joined to overlap the layer below, all the parts being fastened by iron rivets. In order to minimize the longship's weight and thereby maximize its speed, the shipwrights planed the strakes until they were no more than one inch (2.5 cm) thick.

Rowers sat on benches, or, in ships designed for long voyages, on sea chests that contained their belongings and that were designed to fit into the hull. When the side planking reached the appropriate height the shipbuilders fitted ribs and crossbeams inside the hull, with the benches or chests secured to them. Moss soaked in tar was used to make the hull waterproof. The figure carved on the prow may have been meant to strike fear into the sailors' foes, or it may have protected the crew from the fearsome gods and monsters they believed inhabited the depths of the sea.

The rudder was fitted to the side of the ship and the mast was mounted in the center, set securely into a large block of wood. The square, woolen sail hung from a yard, and a pole connected to one of the bottom corners made it possible to turn the sail, thus allowing the ship to tack into the wind. Out at sea a longship relied mainly on the wind for propulsion, but when the crew needed to accelerate or maneuver in an enclosed space, the rowers, seated facing the stern, provided the power.

The Scandinavians were later in introducing masts and sails than many other North Europeans. Some historians believe this may have been due to a tradition that it was lazy to rely on the wind: "Real men row!" Whatever the reason, once they adopted sail power, the longship design proved highly adaptable. The ships varied in size and served many purposes. Not all longships carried warriors intent on pillage, but the warriors were formidable and once they had ships capable of long sea journeys, the Vikings, who had been raiding Scandinavian coastal communities for centuries, began to look farther afield.

The first Viking raid on Britain was on June 8, 793, when the *Anglo-Saxon Chronicle* recorded that "the ravages of heathen men miserably destroyed God's church on Lindisfarne, with plunder and slaughter." One of the Norse *sagas*—Icelandic and Norse histories of kings or important families, often recounting heroic deeds or adventures—tells of an attack on a farm in which the farmer and his family captured the raiders and tied them up. During the night Egill, the Viking leader, escaped from his bonds and released his men. They stole the farmer's goods and headed back to their ship, but on the way Egill felt guilty for having stolen the farmer's property. So the raiders returned to the farmhouse, set it on fire, and killed all the occupants as they tried to escape. The Vikings were then able to return home as heroes who were fully entitled to their booty, rather than as mere thieves.

THE VIKING HAFSKIP

In Old Norse, which is the language the Vikings spoke, *fara í víkingr* meant "to go on an expedition," and a person taking part in an expedition was a *víkingr.* Although the name inevitably conjures images of fierce warriors wreaking havoc on peaceful farming communities, not all *víkingr* were men of violence. Some were farmers who took their families with them on their travels. It was not plunder that they sought, but pasture for their sheep and a place to build a home. They came as settlers. Other *víkingr* were traders with goods for sale and an eye for a bargain.

The ships that carried these more peaceful voyagers were built to the basic longship design, but they were not longships. The Old Norse name for a longship was *langskip,* and the vessel the settlers and traders used was a *hafskip* or *knorr,* also spelled *knarr* and *knörr.* Intended for long sea journeys, the hafskip was much broader than a longship and some hafskips were much larger overall. An average hafskip was about 50 feet (15 m) long and 15 feet (4.5. m) wide, but some were more than 70 feet (21 m) long and up to 20 feet (6 m) wide.

On a longship, the mast could be unstepped when the sail was not required, but the mast was a permanent fixture on a hafskip, and a hafskip had only four to seven pairs of oars, in contrast to the longship's 15 to 20—or even as many as 30—pairs of oars. The hafskip was designed for sailing rather than rowing, and its oars were used only for docking and for keeping the ship facing into the wind during storms. The single square sail was made from woven woolen cloth secured with ropes of sealskin or walrus skin, and it was large, held on a mast that was up to 40 feet (12 m) tall. There was a rudder on the starboard quarter and the hafskip had two boats, one carried onboard and the other towed behind. The ship also carried awnings that could be raised to protect the occupants from hot sunshine or cold rain—but not, of course, from strong winds.

These ships were very seaworthy. One, built in the ninth century and used to bury a Viking king at Gokstad Farm, in Sandar, Norway, was excavated in 1880; an exact replica of it was then built. In 1893 this replica sailed from Bergen to New York, via Newfoundland, in 27 days, then continued up the Hudson River, through the Erie Canal into the Great Lakes, and finally to Chicago. The Gokstad ship, and its replica, carried 32 oars and a sail with an area of about 1,184 square feet (110 m^2) that could have given the ship a top speed of about 12 knots (14 MPH, 22 km/h).

A large hafskip transporting settlers carried at least 30 men, together with their wives and children, cattle, sheep, and dogs, tools and weapons, and enough food and clothing for all the people and animals for a journey of several weeks. If it was on a trading mission a hafskip could carry about 25 tons (23 tonnes) of cargo. This might have consisted of barley or wheat grain, milled barley or wheat flour, furs, woven cloth, or walrus ivory. Viking traders were also in the business of buying and selling slaves.

The hafskips carried Scandinavian settlers to Britain, Ireland, and other parts of Europe, even as far away as southern Italy and Sicily. They colonized Iceland and western Greenland, and they visited North America (see "Eric the Red and Greenland" on pages 47–49 and "Leif Ericson and Vinland" on pages 49–51). But soon after settlers had built dwellings and established farms, and traders were making regular rounds of the settlements, rulers back in Scandina-

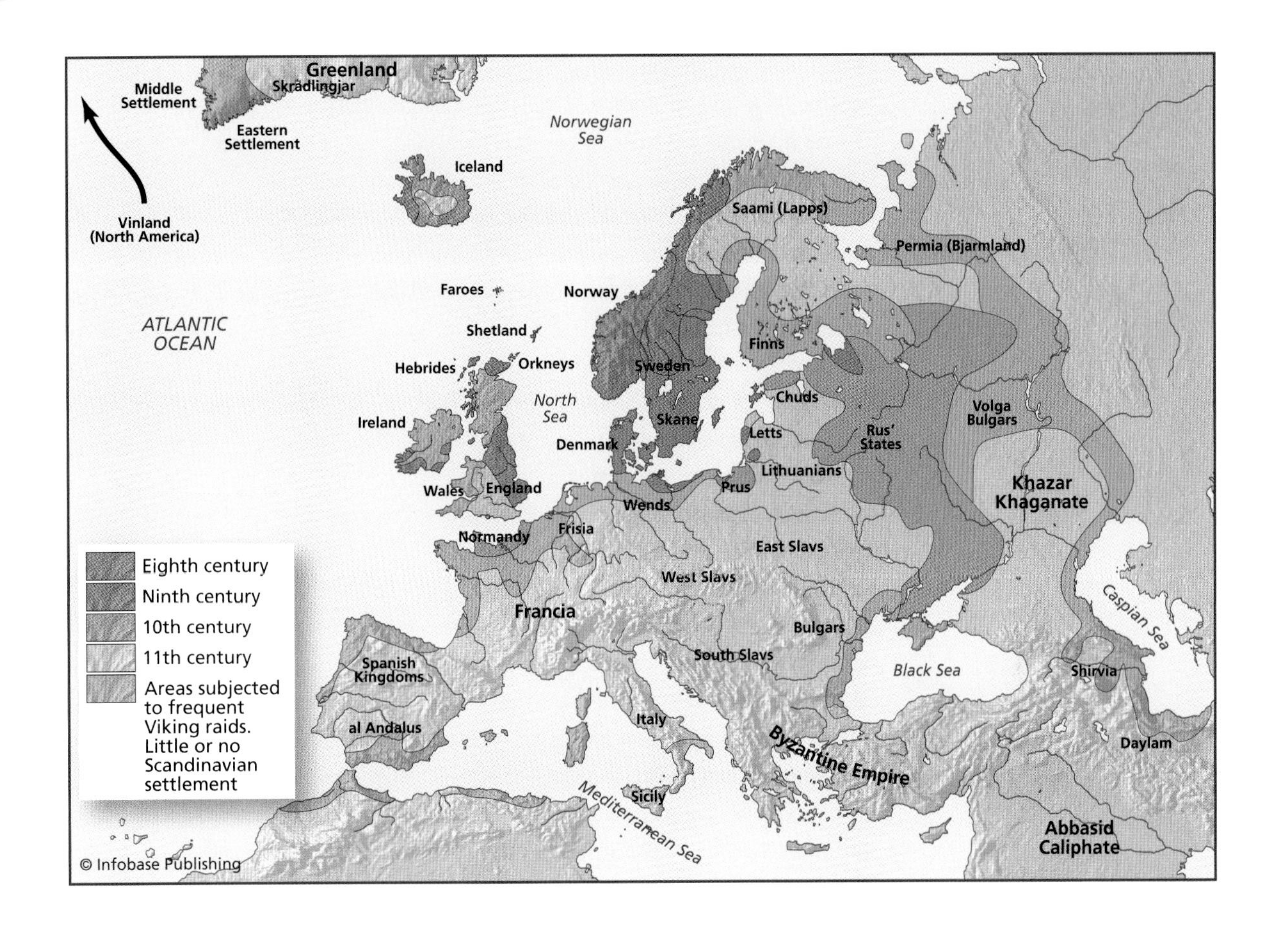

Viking warriors raided coastal communities throughout western Europe, and in places their influence extended far inland. This map shows how their settlements expanded from the eighth century until their final decline in the 11th century.

via began demanding taxes. They had mixed success, depending on the distance to the people they claimed as subjects, but by the 11th century half of England was under Norse rule. The map above shows the area in which Vikings settled, and those parts of Europe and the North African coast that suffered repeated Viking raids but where there was no significant Viking settlement.

KON-TIKI, RA, AND TIGRIS

The peoples of Polynesia, Micronesia, and Oceania are the descendants of seafaring people from Southeast Asia. Their origin, now firmly established, was once a mystery, because the islands of the Pacific are widely scattered over a vast area of ocean and it was difficult for European and American anthropologists to imagine how people could have sailed the distances involved without the benefit of any modern technology—not even the magnetic compass. During time he spent on the Marquesas Islands in 1937 and 1938, the Norwegian explorer Thor Heyerdahl (1914–2002; see sidebar on page 27) observed that in the tropical South Pacific the prevailing winds and ocean currents move from east to west, which meant that travelers from Asia would have had to sail against the winds and currents. Heyerdahl accepted that the ancestors of the islanders had lived in Southeast Asia, but believed they had been carried first to America and then spread across the Pacific from North and South America. He determined to test the feasibility of such a journey in a vessel of the type used by the Inca civilization of Peru. This led to the *Kon-Tiki* expedition.

Kon-Tiki was an old name for the Inca sun god Viracocha. The craft on which the expedition sailed was a raft made from nine balsawood logs up to 45 feet (13.7 m) long and 2 feet (60 cm) in diameter, fastened together by hemp ropes. Smaller logs were fastened across the main logs at three-feet (1-m) intervals to make the structure more rigid, boards were fixed to the bow to give some protection from splashing, and a centerboard was wedged between the logs. A steering oar was fitted at the stern. The mast, made from mangrove wood, was in the form of an A-frame 29 feet (8.8 m) high. The raft had a mainsail measuring 15 by 18 feet (4.5 by 5.5 m); a topsail was carried above the mainsail and there was also a sail mounted on a smaller mast at the stern. The raft had a cabin measuring 14 by 8 by 4–5 feet (4.3 by 2.4 by 1.2–1.5 m) made from plaited bamboo with a banana-leaf thatch. A deck of split bamboo partly covered the logs. No metal was used anywhere in the construction.

The crew consisted of seven men, including Heyerdahl. The illustration on page 28 shows them on the *Kon-Tiki,* waving on their arrival at San Francisco. They took with them 66 gallons (250 l) of water stored in vessels made from bamboo tubes, a variety of fruit and vegetables, canned food, and army field rations. On the crossing they augmented this diet with fish that they caught.

Kon-Tiki sailed from Callau, Peru, on April 28, 1947. Once clear of the coast, it sailed northward with the Peru Current then westward with the South Equatorial Current. The expedition came within sight of Puka-Puka atoll (14.8°S, 138.8°W) on July 30 and reached Angatau Island (15.8°S, 140.9°W) on August 4, but did not land. The voyage ended on August 7, when the raft struck a reef and the crew managed to beach it on an uninhabited island near Raroia Island (16.0°S, 142.4°W). They had sailed approximately 3,770 nautical miles (4,335 miles, 6,975 km).

Heyerdahl led an archaeological expedition to Rapa Nui (formerly Easter Island) in 1955–56. There was a widespread belief—which Heyerdahl promoted—that a race of white people had settled

THOR HEYERDAHL

The Norwegian adventurer, explorer, and ethnographer was born at Larvik on October 6, 1914. His father, also called Thor, was a master brewer and his mother, Alison Lyng, had a keen interest in biology. She inspired and stimulated her son's interest in zoology and anthropology.

Heyerdahl enrolled at the University of Oslo in 1933 to study zoology and geography. While there he became interested in Polynesian history and culture and with the help of his zoology professors, Kristine Bonnevie and Hjalmar Broch, Heyerdahl devised and secured funding for a project to visit the Marquesas Islands. His aim was to study the indigenous animals and try to discover the route by which their ancestors had reached the islands. On December 24, 1936, Heyerdahl married a woman whom he had met before he entered the university. His wife, Liv, traveled with him on almost all of his expeditions.

On December 25 the couple traveled by train to Marseille and from there by sea to Tahiti and then to the Marquesas. This expedition lasted from 1937 to 1938, and his observation that the prevailing winds and ocean currents were from the east led Heyerdahl to doubt the conventional scientific view that the Pacific Islands had been colonized directly from Asia, in the west. He came to believe that although the Polynesian peoples came originally from Southeast Asia, the winds and currents must have carried them first to North America.

Heyerdahl resigned from the University of Oslo following his return from the Marquesas and began to devote his time to investigating the routes by which Polynesians crossed the Pacific. In 1939 and 1940 he studied Native American culture in British Columbia, Canada. This led him to conclude that the earliest wave of Pacific colonizers had crossed the ocean from South America, sailing on balsawood rafts. This led to the 1947 *Kon-Tiki* expedition. Later expeditions took him to the Galápagos Islands in 1952 and Rapa Nui in 1955–56. While in Rapa Nui, Heyerdahl became interested in the papyrus reed boats used by the early Egyptians, and in the feasibility of making ocean crossings in them. This led to the two *Ra* expeditions of 1969–70. The *Tigris* expedition of 1978 crossed the Indian Ocean by reed boat. Heyerdahl studied ancient ruins in the Maldive Islands from 1981 to 1984, returned to Rapa Nui from 1986 to 1988, and from 1988 to 2002 was in Peru. He was also involved in many other archaeological projects.

Heyerdahl always worked with crews drawn from many countries. He was vice-president of the World Association of World Federalists, working to strengthen the influence of the United Nations, and an advisor to the World Wildlife Fund International. In 2002 he retired to his home at Colla Micheri, Italy, where he died from a brain tumor on April 10. His body was sent to Oslo; where on April 26 Heyerdahl received a state funeral in Oslo Cathedral. After his cremation, his ashes were returned to Colla Micheri.

The crew of the *Kon-Tiki* expedition waving to the crowds on their arrival at San Francisco on September 29, 1947. From left, they are: Thor Heyerdahl, Bengt Danielson, Erik Hesselberg, Torstein Raaby, Herman Watzinger, and Knut Haugland. *(Associated Press)*

the island before the Polynesians reached it and that the Polynesian Rapa Nui islanders had destroyed their own culture through infighting and clearing all the forests. In fact, Rapa Nui had very fertile soil and abundant fish, and the island had supported a large population with a highly developed culture. What had really happened was that Europeans captured many of the islanders as slaves, killed more in the process, and brought in Polynesian settlers from other islands; it was European oppression that had destroyed the original culture.

While he was in Rapa Nui, Heyerdahl became interested in the boats made from papyrus reeds that had once been used in Egypt (see "Egyptians on the Nile" on pages 2–4). He believed that the people who sailed from South America to colonize the Pacific Islands inherited their culture from Egypt. This implied that the Egyptians had crossed the Atlantic in reed boats. The *Ra* expeditions of 1969 and 1970 set out to test the feasibility of this theory.

Ra was made from bundles of papyrus reeds. It had a single, inverted V-shaped mast with a single square sail, and a steering oar on the port quarter. The prow and stern curved upward, rising high from the water, and there was a cabin toward the stern. The vessel was built in Chad and transported to Morocco. The expedition sailed from Safi, Morocco, on May 25, 1969. All went well until several bundles of reeds became detached on the starboard side. The stern section started to sink, a hurricane threatened, and on July 18 the *Ra* had to be abandoned. Undeterred, Heyerdahl commissioned another boat to be built using an improved design, and he brought four Aymara boat builders from Lake Titicaca to help with the construction. The Aymara still constructed reed boats using techniques similar to those used in ancient Egypt. This became *Ra II. Ra II* sailed from Safi on May 17, 1970, and reached Barbados on July 12, having covered approximately 3,790 miles (6,100 km) in 57 days. The map below shows the routes the two vessels followed. Heyerdahl had provided no evidence that Egyptians ever crossed the Atlantic, but he had shown that a papyrus vessel was more seaworthy than might have been supposed.

In 1978 Heyerdahl embarked on another expedition in a papyrus boat. This time he aimed to test the idea that the Mesopotamians,

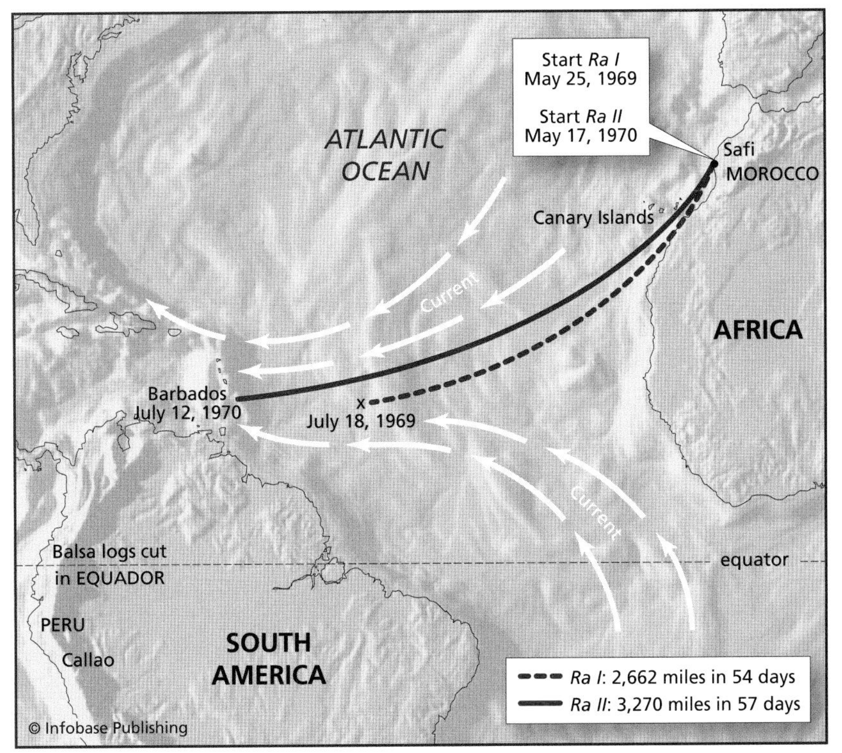

Ra I sailed on the trade winds, but the vessel began to disintegrate after eight weeks, and the voyage had to be abandoned: *Ra II* fared better, sailing from Safi, Morocco, to Barbados.

living in what is now Iraq, might have made contact with the flourishing civilization in the Indus Valley, Pakistan. Mesopotamians would have used reeds from the marshes in southern Iraq to construct their boats, and Heyerdahl commissioned Marsh Arabs, helped by the four Aymara Native Americans, to build the boat he named *Tigris.*

Tigris was Heyerdahl's largest reed boat, 60 feet (18 m) long. It sailed in November 1977, with a crew of 11, down the Shatt-el-Arab waterway and into the Persian Gulf. They called at Bahrain, Muscat in Oman, and then crossed the Gulf of Oman into the Indian Ocean. They reached the mouths of the Indus River, Pakistan, and then followed a more southerly route across the Arabian Sea and Indian Ocean to the Horn of Africa. Heyerdahl had wanted to land at Mitsiwa (now called Massawa) in Ethiopia, but war between Ethiopia and Eritrea made this impossible. To the south, Somalia was also at war and *Tigris* could not put ashore there, either. On the opposite side of the Gulf of Aden, Yemen was at war. The crew did the only thing they could, and sailed down the middle of the Gulf of Aden. Finally they stepped ashore in Djibouti, having spent five months at sea and traveled about 4,225 miles (6,800 km). The map on page 31 shows the route of the *Tigris* voyage.

Heyerdahl and his crew were appalled by the extent of the warfare throughout the region. Although *Tigris* was still seaworthy they decided to destroy it publicly and on April 3, 1978, they burned the ship at Djibouti. *Tigris* had sailed under the flag of the United Nations and on the day it burned Heyerdahl wrote an open letter to the secretary general of the UN. The letter included the following passage:

> We are able to report that in spite of different political views, we have lived and struggled together in perfect understanding and friendship, shoulder to shoulder in cramped quarters through calm and storms, always according to the ideals of the United Nations: cooperation for joint survival. When we embarked last November on our reed-ship *Tigris* we knew we would sink or survive together, and this knowledge united us in friendship. When we now, in April, disperse to our respective homelands, we sincerely respect and feel sympathy for each other's nations. . . .
>
> Today we burn our proud ship, though the sails and rigging are still up and the vessel is in perfect shape, to protest against inhu-

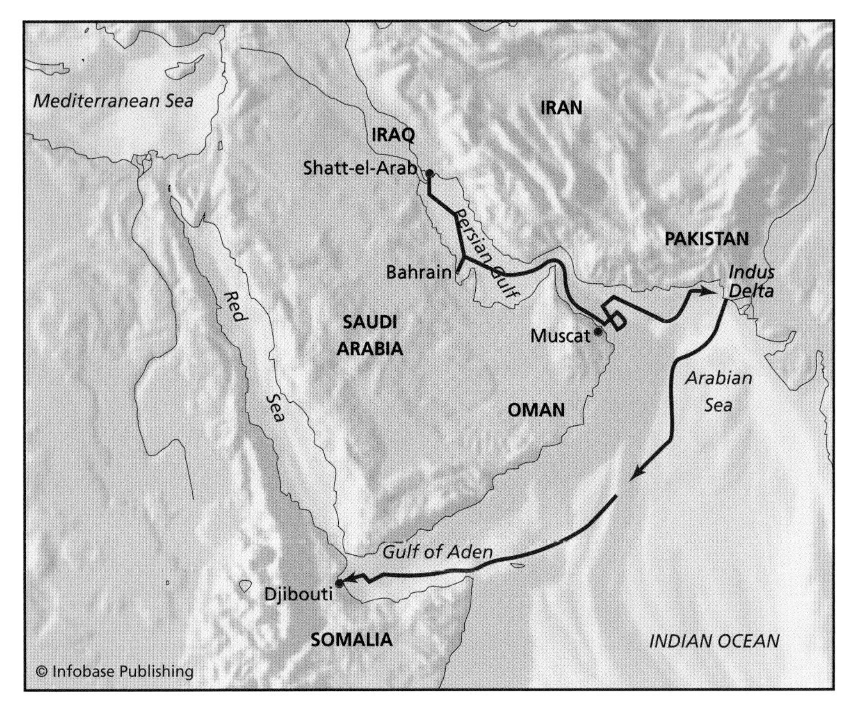

The *Tigris* sailed from the Shatt-el-Arab, Iraq, through the Persian Gulf, and into the Indian Ocean. The ship visited Muscat, Oman, and the mouth of the Indus River, Pakistan, before crossing the ocean to Djibouti, East Africa.

man elements in the world of 1978 to which we have come back as we reach land after sailing the open seas. Now we are forced to stop at the entrance to the Red Sea. Surrounded by military airplanes and warships from the world's most civilized and developed nations, we have been denied permission by friendly governments, for reasons of security, to land anywhere, but in the tiny, and still neutral, Republic of Djibouti. Elsewhere around us, brothers and neighbors are engaged in homicide with means made available to them by those who lead humanity on our joint road into the third millennium. . . .

We are all irresponsible, unless we demand from the responsible decision makers that modern armaments must no longer be made available to people whose former battle axes and swords our ancestors condemned.

Heyerdahl's voyages were spectacular feats of boatbuilding and seamanship. They caught the public imagination, which greatly increased popular interest in archaeology and anthropology, but Heyerdahl's scientific purpose was to substantiate his theories of cultural diffusion. He did not think it possible that Polynesian cultures could have developed without initial stimulation from more advanced civilizations across the ocean. He succeeded in demonstrating the feasibility of making long ocean crossings in vessels of ancient design, but scientists have not accepted his diffusionist theory.

Crossing the World

A person who loves books is known as a *bibliophile.* The French word for library is *bibliothèque.* A list of books is a *bibliography.* The Christian and Judaic scriptures consist of books that are published as a single volume, called the *Bible.* The root of all these words is *bibl* and it refers to Byblos, a city in Lebanon. Originally the city was called Gubla, then Gebal, and today it is called Jubayl or Jbeil. It was the Greeks who called it Byblos, which was the Greek word for papyrus, the sedge that found many uses in ancient times, one of which was to make paper. Byblos is where the Greeks obtained their writing paper, and the people who sold it to them were Canaanites, whom the Greeks called *phoinikèia.* The suppliers came to be known as the Phoenicians.

The Phoenicians were seafarers whose prosperity was founded on buying and selling, rather than military conquest. It was not lands they sought, but markets and goods to supply them. This chapter tells of peoples who crossed oceans for peaceful purposes. The Phoenicians were traders. Other travelers were searching for places to live, places where food was abundant and reliable and where they could live in safety. They were chasing their dreams—and finding them.

THE PHOENICIANS AND THE MEDITERRANEAN

In about 3200 B.C.E. a small town on the eastern coast of the Mediterranean that had existed since 4500 B.C.E. began to expand rapidly, growing into the city of Byblos. By 3000 B.C.E. Byblos had civic buildings, several temples, and a strong city wall protected it on the landward side. Byblos was in the land of Canaan and its inhabitants called themselves Kena'ani (Canaanites), but they were developing a culture that distinguished them from other Canaanites. They were becoming the people the Greeks knew as the Phoenicians. Their country, Phoenicia, lay in what is now Lebanon together with parts of Israel, Syria, and the Palestinian Territories. Phoenicia consisted of several city-states, the most important of which were Byblos, Tyre, and Sidon. The Egyptians regarded the cedar of Lebanon as the source of the best timber for shipbuilding. They obtained it from Byblos and in about 1500 B.C.E. they conquered Phoenicia. The Egyptians were the first of many occupying powers that were to impose their rule on the Phoenicians.

Repeated invasion and occupation might have destroyed the Phoenicians completely, but they possessed talents with which they changed the world. The Phoenicians were remarkably skilled shipbuilders and seafarers, they used their ships principally to trade, and they used an alphabet to write their language. The eastern Mediterranean has always been a region where European, African, Egyptian, and Asian cultures have met and mingled. The Phoenicians were familiar with all of them and learned from all of them, but adapted what they learned to their own needs.

Most Phoenician ships were merchantmen, built to a design that they perfected early and subsequently changed little. Merchant ships were known as round boats because of their shape; the illustration on page 35 shows a model of a typical example. These ships were very broad, to maximize the amount of cargo they could carry. They had a high stem and stern and a carved prow, typically in the shape of a horse's head. There was a central mast with a square sail that was strengthened with strips of leather. Phoenician ships had oars, but on the open sea they relied mainly on the sail. The large rock at the bow was an anchor. Earlier versions of this design had two steering oars, but in this model there is a single oar that passes through the stern, and thus is a primitive rudder. Phoenician merchant ships also carried a lifeboat, although historians are uncertain whether this was carried on board or towed behind.

In ships like this the Phoenicians visited all the important coastal cities around the Mediterranean, carrying goods of all kinds. Their earliest export was timber from the cedar of Lebanon forests close

A model of the type of Phoenician ship that sailed the Mediterranean in about 900 B.C.E. *(Scala/Art Resource, NY)*

to Byblos. They made the logs into rafts that they towed behind their ships. The Phoenicians living at Tyre also collected a local sea snail, *Haustellum brandaris,* which produced a clear secretion that turned into a purple dye upon exposure to air. The dye was known as Tyrian purple, royal purple, or imperial purple, and it was highly prized by the Greeks and later by the Romans.

As their seamanship improved and their confidence grew, the Phoenician sailors moved farther from the coast. They learned to steer by the stars, and it may have been the Phoenicians who divided a circle into 360 degrees and named the points of the compass. They may also have discovered how to measure a ship's speed. Eventually they began venturing beyond the Mediterranean. They sailed southward along the West African coast as far as Senegal, and northward along the Iberian coast until they entered the Bay of Biscay, notorious for its heavy seas. They sailed beyond Biscay and reached the southwestern peninsula of Britain. The map on page 36 shows Phoenicia and the trade routes the Phoenicians developed.

Although the Phoenicians traveled in peace, other nations were more warlike and the Phoenicians needed to defend themselves, so

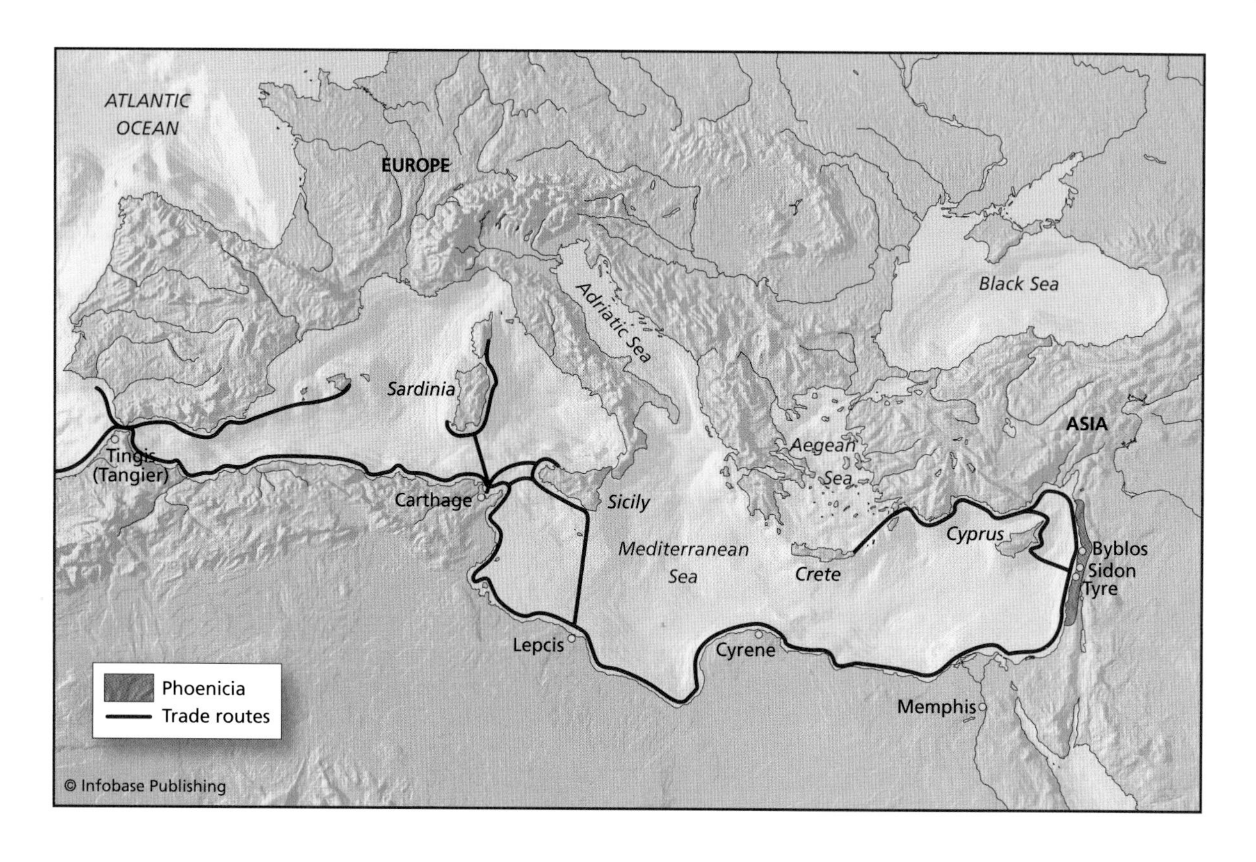

Phoenicia was a small country, but its merchants were known far and wide. By establishing trade links among ports around the Mediterranean, the Phoenicians helped ancient economies to flourish, but as well as goods and money, the merchants brought news, gossip, and ideas.

they also built warships. These were narrower than the merchant ships. They were galleys, relying on their oars more than their sails. They had very large oars at the bow and stern; and these were used to turn the ship quickly. When the bow oar was not in use it could be fastened to the prow and used as a battering ram. By about 700 B.C.E. the Phoenicians were using warships with two banks of oars—biremes—and a ram in the form of a sharp beak at water level.

All merchants and ship owners need to keep records of clients, orders, suppliers, purchases, and sales, as well as accounts of money transactions. Those records and accounts have to be written down. The Egyptians wrote in a hieroglyphic script and some other cultures used versions of the Sumerian cuneiform script, which involved pressing a wedge-shaped stylus into soft clay. Both methods were very complicated and difficult to learn. Consequently, only a few people were able to read and write. The Egyptian system was partly phonetic, however, with symbols standing for sounds rather than whole words, and this gave rise to a set of symbols that were used to write the languages spoken by Canaanite peoples. The Phoenicians adapted those symbols to the needs of their own language.

Everywhere they went, the Phoenician merchants used their system of writing to keep their records, and everywhere they went their customers and suppliers watched them doing it. The Phoenician alphabet had 22 letters but no vowels, and the language was written from right to left. People who could read the script would have known how the sequence of consonants should be pronounced. The illustration below shows the Phoenician alphabet.

People the length and breadth of the Mediterranean world adopted the Phoenician alphabet, then modified it to suit their own languages. Arabic and Hebrew scripts are descended from the Aramaic script,

The Phoenician alphabet had 22 characters, with no vowels. Phoenicians wrote from right to left.

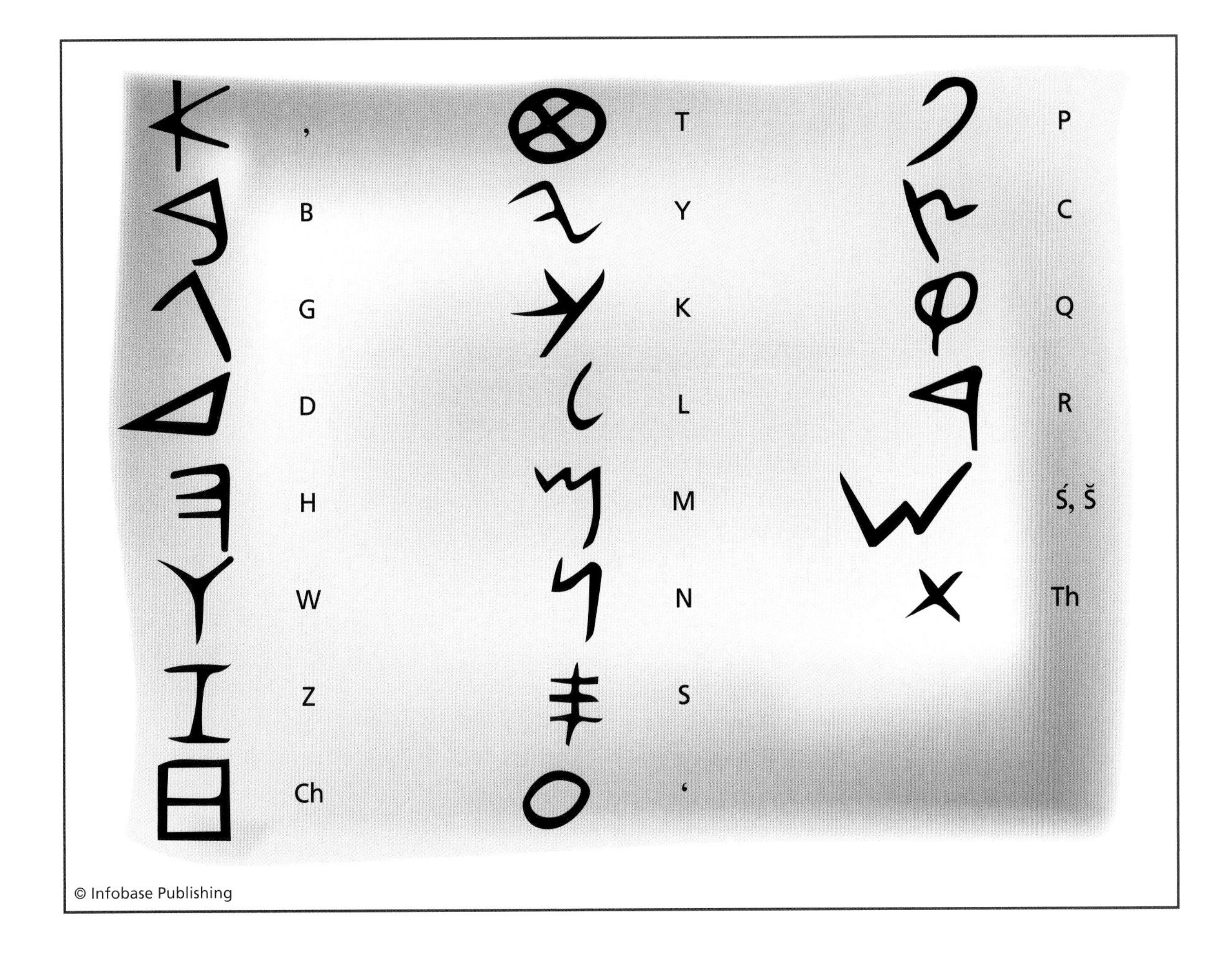

which is a version of Phoenician, and the Greek, Latin, Cyrillic, and Coptic alphabets are direct descendants of Phoenician.

COLONIZING THE PACIFIC ISLANDS

The Pacific Ocean is dotted with thousands of islands. Most of the islands are inhabited, but vast expanses of ocean lie between them. They lie in three areas: Polynesia, Micronesia, and Melanesia. The illustration on page 39 shows these three divisions, and the islands and island groups that are included in them.

The people who settled them were seafarers from Southeast Asia, who had set out from Taiwan. Today, most Taiwanese people are Chinese, but the Chinese did not arrive in Taiwan until the 16th century. It was the earlier inhabitants, who had lived there for about 15,000 years and whose descendants now account for about 2 percent of the Taiwanese population, who were the ancestors of modern Polynesians and Micronesians. Those early wanderers sailed from Taiwan through the islands of Indonesia, and in about 1000 B.C.E. they reached Melanesia. The islands of Melanesia were already occupied by small, isolated groups of people who had arrived very much earlier (see "Arrival in Australia" on pages 41–42). The Asian wanderers appear to have had only limited contact with the Melanesians. They sailed among the islands, but then they left, heading into the ocean, eventually to Micronesia and Polynesia.

This theory of the origins of Polynesians is based on studies of mitochondrial DNA (mtDNA), which a person inherits exclusively from his or her mother. These show a clear matrilineal relationship between Polynesians, Micronesians, and the original inhabitants of Taiwan. An alternative theory, however, based on studies of the Y chromosome, which is passed on from father to son, suggests that Polynesian males are descended from Melanesian males. It is possible, therefore, that Polynesians are descended from both Taiwanese and Melanesian peoples. However, the matter remains undecided.

In about 3000 B.C.E. people living in the Bismarck Archipelago in Melanesia began making distinctive pottery. These people were most likely ethnic Chinese or Taiwanese, and in addition to making pottery they kept domesticated dogs, pigs, and chickens, and grew vegetables. Theirs is known as the *Lapita culture* and in about 1300 B.C.E. it began to spread eastward through the migration of people from

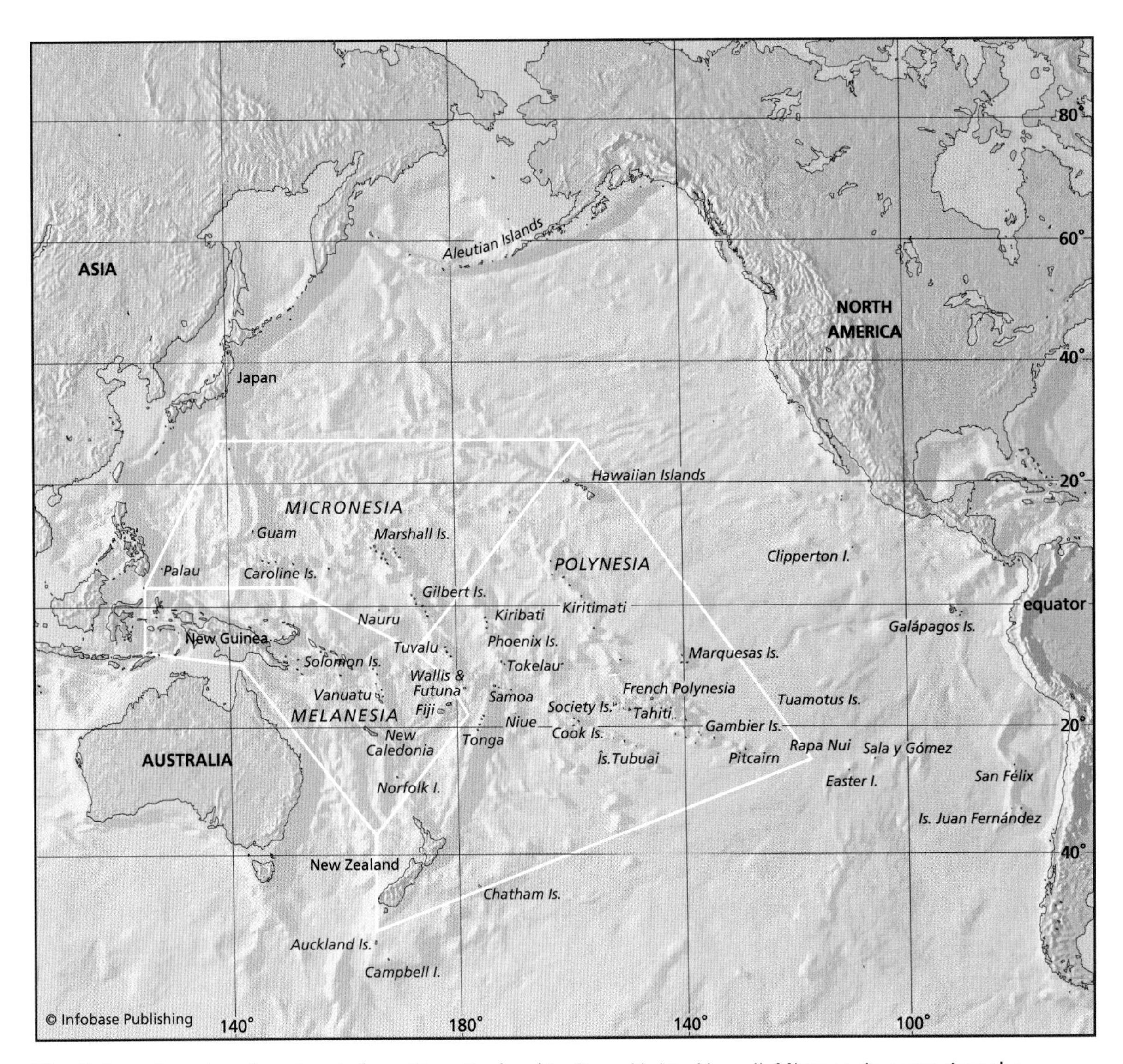

The Polynesian triangle extends from New Zealand to Rapa Nui to Hawaii. Micronesia comprises the islands to the north of New Guinea. Melanesia comprises New Guinea and the adjacent islands.

the Bismarck Islands. The culture reached Fiji, Tonga, and Samoa—a region 3,700 miles (6,000 km) distant—in about 900 B.C.E. The people following this way of life settled in the region for a time, but in about 300 B.C.E. their descendants resumed their eastward migration. They migrated to the Cook Islands, Tuamotus, and the Marquesas; from there, they migrated yet again, this time to Rapa Nui and Hawaii, settling there in about 500 C.E. Finally, in about 1000 C.E., Polynesian settlers—the Māori people—arrived in New Zealand.

The Polynesians sailed outrigger canoes (see "Outriggers" on pages 4–6). The canoes used for long voyages were approximately 65 feet (20 m) long. They carried a crew of at least five and up to 15, with a store of vegetables, live chickens and pigs, and water stored in gourds that could be augmented by collecting rainwater in sailcloth. When James Cook's expedition reached Tahiti in 1774, Cook and his officers met Polynesians and learned their language, and they managed to persuade a Tahitian called Tupaia to tell the story of his people's colonization of the Pacific Islands and how it was achieved. Tupaia explained how Polynesian seafarers were able to navigate when they were out of sight of land.

Navigators memorized the routes to other islands and taught them to young men who were learning the skill, in the form of songs and drawings. Their method was based first on their knowledge of the direction and times of rising and setting of the most prominent stars and planets. Voyages began at dusk. The navigator set a course in relation to the direction of prominent landmarks that were still visible and of the stars, and during the night he would steer by the stars. During the day he would steer by the Sun. The condition of the sea and direction of the wind also provided valuable information. In the tropical Pacific, the prevailing winds blow from the northeast to the north of the equator and from the southeast to the south of the equator. Because they blow for most of the time, the winds produce a large swell, with waves that all move in the same direction. Navigators could steer by the direction of their canoe in relation to the swell, and they would tow a length of rope in the water behind the canoe. If a sudden wave or gust of wind blew the canoe to one side, the rope would not be affected and its line would record the direction they should steer. Small pennants tied to the rigging and mast indicated the wind direction, which was also useful. In addition, Polynesian sailors knew the paths followed by migrating birds and whales. They were familiar with ocean currents, the cloud patterns that formed over distant islands, and they watched for rafts of plant debris that indicated land just over the horizon.

Polynesian navigation was remarkably skillful—but why were the Polynesians so restless? One possible clue can be found in the fact that despite living on islands surrounded by abundant stocks of fish, the people of the Cook Islands seldom eat fish. The seas around Rapa Nui support many species of edible fish, as well as lobsters and turtles. These have always been an important part of the islanders' diet and island traditions forbade fishing at certain times of year, which prevented overexploitation of the stocks. As recently as the early 20th century, however, the islanders believed that the fish living in deep water farther from shore were poisonous and they refused to touch them.

In 2009, Teina Rongo, a Cook Island marine biologist studying for a Ph.D. at the Florida Institute of Technology, aroused considerable interest among historians when he proposed that from time to time the fish on which Polynesians depended had turned poisonous and the islanders had suffered from a type of food poisoning called *ciguatera.* Robbed of their food supply, they had had no alternative but to move elsewhere. Ciguatera is caused by eating fish contaminated by a single-celled organism called *Gambierdiscus toxicus.* The poisoning is seldom fatal, although it can be; its symptoms include vomiting, diarrhea, abdominal cramps, pain in many parts of the body, blurred vision, tingling sensations, a burning sensation on contact with a cold surface, and heartbeat irregularities. Common throughout the Tropics, this poisoning is most severe in people who have eaten carnivorous reef fish such as barracuda, snapper, and grouper. It may be that the waves of Polynesian migration were motivated by the need to find wholesome food.

ARRIVAL IN AUSTRALIA

On February 26, 1974, Jim Bowler, a geomorphologist at the University of Melbourne, discovered human remains among the shifting sand dunes near Lake Mungo, a dry lake in New South Wales. Scientists quickly agreed that these were the oldest remains of an anatomically modern human ever found in Australia, but it was not until 2003 that they were able to determine just how old they were. Mungo Man, as he was called (or, less romantically, Lake Mungo 3), had lived about 40,000 years ago. He had been a large man, six feet five inches (1.96 m) tall, and he had been buried ceremoniously, with his body sprinkled with red ochre. Not far away, scientists also found stone tools that were 50,000 years old. By 50,000 years ago, humans were living in Australia.

The first settlers reached Australia during the most recent ice age. At that time the ice sheets extending across large areas of Europe and North America held so much water that the average sea level was at least 500 feet (150 m) lower than it is today, and coasts extended much farther than they do now. Australia and New Guinea were joined, forming a single continent geologists have named Sahul; the islands of Indonesia formed a single landmass that was joined to the Malaysian Peninsula; and there was dry land between mainland Australia and Tasmania. Travelers could have walked from Asia to Indonesia, but no farther, because they would have encountered an expanse of deep ocean, dotted with islands, that left Sahul isolated. Tribes moved southward through southern Asia and Indonesia, then crossed from island to island until they found themselves on the continent of Sahul. People had established themselves in Australia by 50,000 years ago, or possibly somewhat earlier, and they continued to migrate southward, reaching Tasmania by about 30,000 years ago.

As they expanded to colonize the continent, the Australians formed themselves into approximately 250 nations, each nation consisting of from as few as five to as many as 40 clans. Each nation had its own language, so at one time about 250 languages were being spoken. Most of those languages have since disappeared or are close to extinction. All of the people lived by hunting game and gathering wild plant foods.

Mitochondrial DNA evidence shows that Aboriginal Australians are closely related to the people of New Guinea and Melanesia. This evidence also suggests that once this part of the world had been colonized, its populations became isolated (although there is some debate as to whether colonization occurred as a single event or in two waves). The Aborigines entered Sahul from Java, but many scholars believe they were descended from the first modern humans *(Homo sapiens)* to have left Africa (see "The Long Walk out of Africa" on pages 217–219). If this theory is correct, then the first Australians were direct descendants of African migrants.

CROSSING THE BERING STRAIT

The western coast of Alaska and Yukon and the northeastern coast of Siberia are separated by a stretch of open sea about 53 miles (85 km) wide. This is the Bering Strait. The sea there is shallow, as are the

Chukchi Sea to the north of the strait and the Bering Sea to the south. During ice ages, when large amounts of water were held in ice sheets covering the northern continents, sea levels fell and so from time to time the Bering Strait was dry land, forming a land bridge between Asia and North America that was 1,000 miles (1,600 km) wide. Scientists call the exposed land Beringia.

Beringia existed for much of the coldest part of the most recent (Wisconsinian) ice age and although it had a cold climate, the surface was not buried beneath ice. Between about 29,500 and 11,500 years ago the Beringian climate was dry and there was a *tundra* type of vegetation consisting of clumps of grass, herbs that flowered and set seed during the very short summer, and dwarf willow. The people living in Beringia were Asians from Siberia, hunters following the caribou and seals, gathering shellfish, and fishing. They occupied Beringia, at least intermittently, but the Cordilleran ice sheet blocked their entrance to North America. As the map on page 44 shows, eastern and northern Alaska lay beneath the ice.

At some time about 14,500 years ago the ice retreated sufficiently to allow access to the Pacific coast of the continent, and entry to the south of the ice sheet. The map shows an ice-free corridor between the Cordilleran and Laurentide ice sheets; archaeologists used to think that corridor provided the route colonists used. Scientists now know, however, that ice closed the corridor until about 11,500 years ago and that the people who settled in Alberta moved there from the south.

The first people to move out of Alaska and into Pacific coastal areas of North America belonged to a group that had been isolated in Beringia for perhaps as long as 15,000 years. Their descendants then spread rapidly across the continent. There is archaeological evidence that people were living in North America 14,300 years ago. Little is known about them, but they probably lived in small, scattered bands, hunting game, fishing, and gathering wild plants.

In 1932 archaeologists excavating a site identified in 1929 near the city of Clovis, New Mexico, discovered stone tools of a distinctive bifacial, fluted type. Similar tools were subsequently found in many parts of North and Central America, and at a few sites in South America, and the people who made and used them were described as the Clovis culture. The Clovis culture lasted from about 13,300 to 12,800 years ago. Their origin is a mystery; they left no traces in Alaska or Canada, and no tools of the Clovis type have ever been

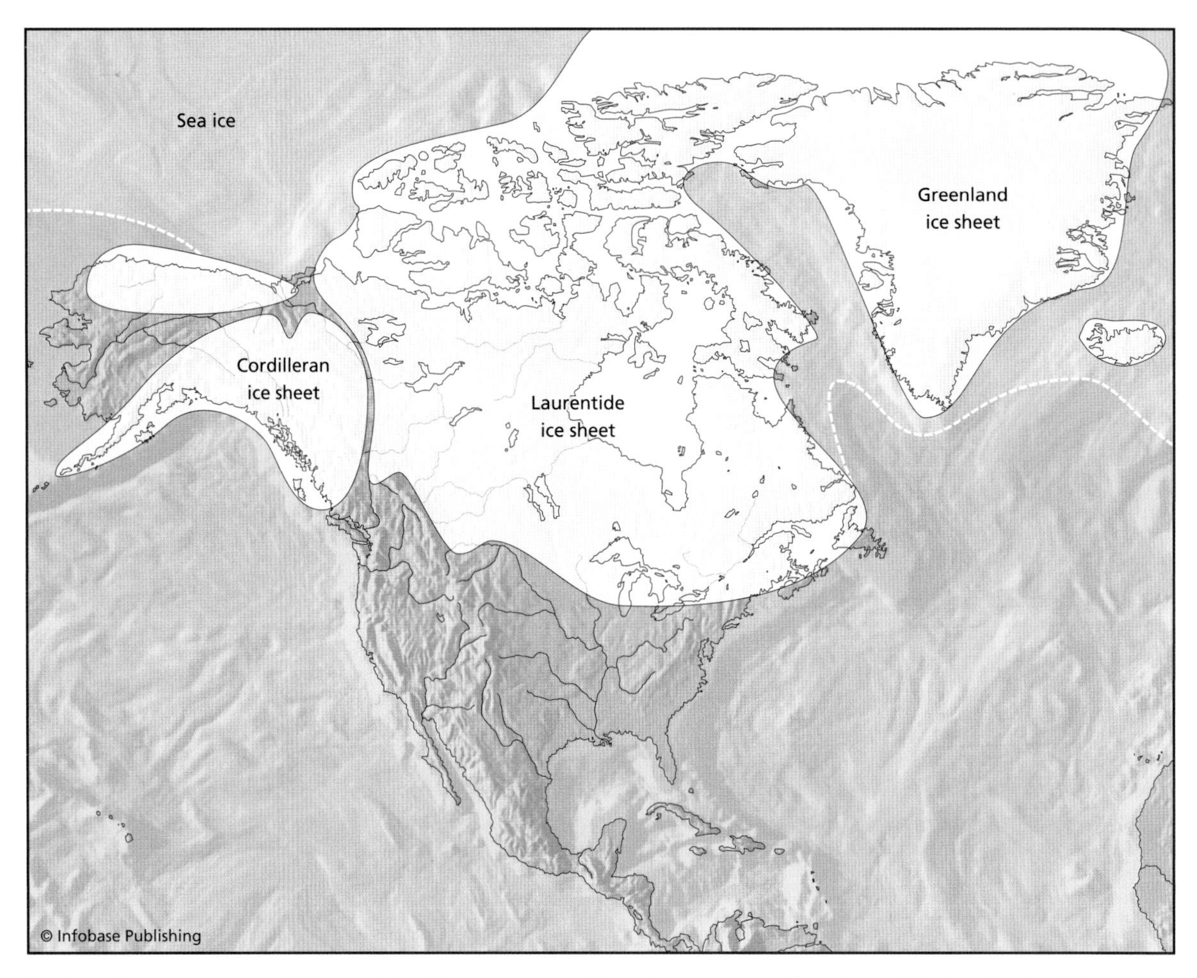

During the most recent (Wisconsinian) ice age, a small ice sheet covered northern Alaska, the Cordilleran ice sheet covered southern Alaska and Yukon, and the much larger Laurentide ice sheet covered the remainder of Canada and the northern United States.

found in Siberia. It is possible that people already living in North America developed the Clovis technology and their culture spread throughout the continent. Be that as it may, the Clovis culture disappeared abruptly 10,500 years ago, to be replaced by a large number of different cultures scattered over the continent. It was about 10,500 years ago that the corridor opened between the Cordilleran and Laurentian ice sheets as the ice sheets were retreating, so perhaps more people moved through the corridor onto the plains and displaced the Clovis culture. So far, no one knows.

Although the Clovis culture was widespread, it was not alone. In 1975 erosion at Monte Verde, Chile, exposed a bone from a mastodont (*Mammut* species) and in 1977 work began on excavating the site. The scientists discovered remains of a structure 20 feet (6 m) long and made from animal skins and wood, with wooden walls, that had been occupied by between 20 and 30 people who lived there sometime between 14,800 and 13,800 years ago. There were hearths, spilled seeds and nuts, stone tools, and traces of 45 different species of edible plants. These people were living in Chile more than 1,000 years before the earliest known Clovis site.

Little by little, scholars are piecing together a coherent account of the way people colonized North, Central, and South America. The most likely sequence is that people first moved from southern Alaska along the Pacific coast, following game herds, fishing, and gathering shellfish. Once south of the ice sheet they moved eastward and when a corridor opened between the Cordilleran and Laurentide ice sheets they moved northward, probably following herds of game. Some of these early settlers may have become the Clovis people, only for their culture to be replaced after a few centuries by new waves of immigrants.

People may also have reached America by sea. Domesticated sweet potatoes *(Ipomomea batatas)* originated in South and Central America. Spanish and Portuguese explorers introduced them to Europe and Africa. However, the inhabitants of several Polynesian islands appear to have been cultivating and eating them long before Europeans reached the New World. It is possible that sweet potatoes may have drifted across the Pacific on rafts of vegetation carried by the prevailing winds and currents. It is also possible that humans transported them and, given their seafaring expertise, those humans were most probably Polynesian—but even if they visited what became America, the Polynesians did not settle there.

OTHAR OF HELGELAND AND THE NORTH CAPE

To medieval scholars the walrus was a fabled beast. Little was known about it, simply because very few explorers had ever ventured far enough north to encounter it. It was known, therefore, mainly by rumor. Albertus Magnus (ca. 1200–80), who wrote a commentary on Aristotle's *Natural History,* was one of the first to mention the walrus, describing it as a kind of whale. If the identity of the animal itself was a puzzle, however, its teeth were known to be a source of valuable ivory, and it was the search for ivory that drew the first Arctic explorer northward.

That explorer's name was Othar, sometimes spelled Ottar, Othere, Ohthere, or Ohthar. Almost nothing is known about him, other than the fact that he was a Norse seaman whose home was in the most northern part of Helgeland, a region of western Norway immediately to the south of the Arctic Circle, and that he set sail for the far north at some time between 870 and 880 C.E. He was described as being a nobleman and landowner of great wealth, influence, and integrity.

Northern Norway at that time was a wasteland, almost deserted except for a few Finnish people who hunted game in winter and fished in the summer. The first purpose of Othar's voyage was to discover how much land lay to the north of this wasteland and whether anyone lived there. He sailed northward for three days, with the land on his right hand and the wide sea on his left, until he was as far north as the whale-hunters ever traveled, and then he continued for three more days. There, he found that the land curved to the east and he had to wait for a westerly wind. When it came the wind carried him for four more days.

Othar had rounded the North Cape, at the northernmost part of Norway, and had continued eastward and then turned south, following the coast until he reached the White Sea. He told of the people he met. At first they were Finns, but he called those living beside the White Sea Biarmians. The Biarmians, who spoke a language very similar to Finnish, told him stories about their country and its neighbors, but Othar pointed out that he was unable to verify these accounts. The Russians called these people Permiaki, or Permians, and the region where they lived was called Permia or Perm.

Apart from his quest for knowledge, Othar was in pursuit of walrus ivory. He knew walruses as *horsevael,* or "horse-whales." The Norwegian name is *hwalros,* "whale-horse," which is the origin of the English name walrus. Othar also knew that their teeth were of great value.

After his return, Othar visited the court of King Alfred the Great (848 or 849–899) and described his voyage to the king. Alfred later mentioned the event in his geographical and historical account of Europe, which was an Anglo-Saxon translation of the historian Orosius with his own additions. Othar made a second voyage some years later, to Ireland and Scotland.

ERIC THE RED AND GREENLAND

The *Landnámabók* (Book of the Settlements), written in Iceland in about 1125, recounts the establishment of Norse settlements in Iceland and it also describes the Norse settlers who went to live in Greenland between about 985 and 1000 C.E. The Vikings had been harrying coastal communities farther south since the latter part of the eighth century (see "Longships" on pages 000–000), but they were unable to explore the north until the climate grew warmer. Before the Vikings began their expeditions there were Irish monks who ventured northward in search of peaceful places to live contemplative lives. One of these, called Dicuil, wrote an account of his experiences in northern seas in the 790s (*Liber de Mensura Orbis Terrae,* Book of measurements of the terrestrial globe). Dicuil stated that the sea was frozen one day's sailing north of Iceland. In 865 a party from Norway led by Floki Vilgerdason tried to start a farm in the country they knew as Snæland (Snowland), but the winter was so harsh that all their cattle died. They noted that Arnarfjord in the northwest of the country was blocked by ice. It was Floki who renamed this hostile country Iceland. Ice is seldom mentioned in later stories, however. The climate was growing warmer and the sea ice was retreating.

One story in the *Landnámabók* provides clear evidence that the sea was warmer than it is today. Thorkel Farserk, one of the earliest settlers in Greenland, wished to entertain his cousin Erik the Red (ca. 950–1003). This meant killing a sheep for the feast, but they were all on the island of Hvalsey. Thorkel had no serviceable boat, so he swam across the Hvalseyjarfjord to the island, caught a sheep, and swam back with the animal. He swam a total distance of more than two miles (3 km). Thorkel may have been very fit, used to swimming long distances, and with an ample layer of body fat to keep him warm, but unless the water temperature was at least 50°F (10°C) he would inevitably have died from cold. Today the average temperature in those waters is 37–43°F (3–6°C). There is also irrefutable evidence of a warmer climate: the Norse settlers buried their dead in ground that is now permanently frozen.

It was in a warmer climate than that of today, therefore, that Erik the Red left his home. Erik Thorvaldsson (Erik son of Thorvald) was nicknamed Erik the Red probably because he had red hair. Erik was born in Norway, but Thorvald committed offences for which he was exiled and he settled with his family in western Iceland. Erik married Thjordhildr (Þjóðhildr) and received land, probably as part of her dowry. But soon he found himself in trouble. Some of his slaves accidentally triggered a landslide that buried the house of Valthiof and his family. Valthiof's kinsman Eyjolf responded by killing the slaves and Erik retaliated by killing Eyjolf. On another occasion, Erik went away to an island, leaving Thorgest to take care of his chair posts—carved poles, usually family heirlooms and extremely valuable, which were erected on either side of their owner's chair. When Erik returned Thorgest refused to part with the poles, so Erik stole them. Thorgest set off in pursuit, and Erik killed Thorgest's sons. Erik was convicted of these two killings in 981 and sentenced to three years in exile.

Erik acquired a ship, probably about 100 feet (30 m) long and in 982, with his family, servants, and livestock, he set sail for the mountains that were visible 175 miles (282 km) to the west from the tops of the Icelandic mountains. Drifting ice prevented them from landing on the coast that was visible from Iceland, so the party continued around the southern tip of the country until they found a place on the western coast where there was pasture for their animals and no ice. Erik and his household farmed the land successfully and Erik spent much of his time exploring. During this time they met no other people, but they did find traces of earlier, Irish settlements.

In 985 Erik returned to Iceland full of stories about the opportunities to be found in this new land, which he called Greenland to make it sound more attractive than Iceland. By the following spring Erik had persuaded about 500 people to sail for Greenland in about 25 boats. Some of the boats turned back and some were lost, but 14 arrived, bringing about 350 people together with their livestock and all their possessions. They established an Eastern Settlement *(Eystribygð)* and later a Western Settlement *(Vestribygð),* and Erik became their paramount chief.

The Greenland settlements grew; by the year 1000 there were probably about 1,000 inhabitants. They were prospering and more settlers were continuing to arrive. The Eastern Settlement had about 190 small farms, 12 churches, a cathedral, a monastery and a convent. The Western Settlement had 90 farms and four churches, and a Middle Settlement—probably part of the Western Settlement—had 20 farms. But in 1002 there was a disease epidemic, probably carried on one of the migrant ships, that caused many deaths. Erik died in 1003, one of the victims.

LEIF ERIKSON AND VINLAND

Erik the Red and his wife Thjordhildr (þjóðhildr) had three sons and one daughter. They called their third son Leif. In later life he was sometimes called Leifr the Lucky. Leif Erikson was born in Iceland in about 970 and sailed with his parents to Greenland. His life was recorded in the Norse sagas, but in accounts that differ at several points. What does seem certain is that Leif sailed to Norway and served at the court of King Olav Trygvasson (ca. 960–1000). The Saga of Erik the Red states that Leif made this journey in 999.

King Olav was a convert to Christianity and established the first cathedral in Norway. During the year Leif spent at the Norwegian court the king converted him; at the end of the year the king sent Leif home to Greenland with instructions to bring Christianity to the Norse settlers. According to the Greenland Saga, Leif returned safely to Greenland. He failed to convert Erik to the new religion, but Thjordhildr did convert; it was she who built the first church in Greenland, at the family estate at Brattahlzð. Leif remained for some time in Greenland, but eventually he grew restless.

An Icelandic trader, Bjarni Herjulfsson, had described how in 986 his ship had been blown off course during a storm and when the skies cleared he had seen a mountainous, heavily forested land far to the west. Determined to find this land, Leif purchased Herjulfsson's boat and equipped it for a long voyage. Leif made landfall in three places. The first was at what he called Helluland (Flat-Stones Land). This was probably on the coast of Baffin Island or northern Labrador. He went ashore again at Markland (Forest Land), which may have been in southern Labrador. His third landfall was at the place he called Vinland (Wine Land).

Although Vinland was mentioned in both the Saga of the Greenlanders and the Saga of Erik the Red, historians regarded references to it as fictitious until 1960, when Helge Ingstad, a Norwegian explorer and writer, discovered traces of a Viking settlement at

L'Anse aux Meadows, on the most northerly tip of Newfoundland's Northern Peninsula. Anne Stine Ingstad, Helge's wife and a professional archaeologist, excavated the site from 1961 until 1968. Her team found eight buildings and a number of artifacts, all of the type being made in Iceland and Greenland in the 11th century. Further excavations took place from 1973 to 1976 under the auspices of Parks Canada. These uncovered three layers of occupation: The lower and upper layers were Native American, and the middle layer was Norse. Some historians believe the L'Anse aux Meadows site is the Vinland described in the sagas, while others believe Vinland was farther south, in New England; other Canadian locations have also been proposed. Regardless of which site corresponds to Vinland, the L'Anse aux Meadows site is definite proof that Norse settlers reached North America centuries before Columbus.

Critics of the L'Anse aux Meadows interpretation point out that wild grapes do not grow in Newfoundland, so it can hardly be called a "wine land." There are several possible explanations for this, however. The first is that other berries, including blueberries, do grow there and can be used to make wine, and blueberries resemble certain varieties of grapes—although they do not grow on vines. A second possibility is that wild grapes did grow there during the Medieval Warm Period, when temperatures were higher than those of today. A third possibility is that although *Vínland* means "wine land," and *vín,* with a long *í,* is the word used in the Saga of the Greenlanders, *Vinland,* with a short *i,* means "meadow land." The saga was written down long after the event, and the author may have made a simple mistake.

Historians now believe that the Vikings made two separate attempts to establish colonies in North America. Neither of these succeeded. Settlers quarrelled among themselves and the Vikings also fought against the indigenous people that they called *skrœlingar,* the word they also used to describe the ancestors of modern Inuit people that they met in Greenland. The North American *skrœlingar* were obviously Native Americans, but it is impossible to identify them more specifically.

The Vikings abandoned their attempts to settle in North America, but much later, in the first half of the 14th century, Greenland Vikings resumed their voyages to Markland. These were to obtain timber, always a precious commodity in Greenland and Iceland. The Greenland settlements had always depended on imports from

Norway of certain essentials, especially iron and timber. By the 14th century, the Medieval Warm Period had given way to the start of the Little Ice Age. The sea ice was extending southward and it was no longer possible for the Viking ships to sail from Iceland to Norway by their traditional route along latitude 65°N. They adopted a more southerly route, but eventually that, too, was closed by ice. They began importing timber from Markland when their contact with Norway was almost completely severed. In 1347 a Viking ship arrived in Iceland with timber from Markland and the crew were arrested for trading illegally and taken to Norway to be tried. Eventually, the sea ice isolated the Greenland settlements completely and the inhabitants died out.

Trading by Desert and by Sea

Trade has always been the most important force driving exploration. Empires often expand by military conquest, but imperial powers do not seek new territories simply for their own sake. Rather, they need markets for their exports and access to the resources of the lands their armies subdue.

This chapter outlines the role of trade in opening routes to distant parts of the world. Many of these routes crossed the oceans, bringing Asian fabrics, spices, ivory, and peacocks to delight wealthy Europeans. Tin, mined in Cornwall, traveled to the Mediterranean lands to be alloyed with Cypriot copper to make bronze. Rome imported food from many parts of its empire.

Where there were ships carrying valuable cargoes there were often pirates eager to seize them. The chapter tells of sea rovers and of legitimate privateers who turned to piracy, and of the careers and fates of some of the most notorious pirates.

Not all trade was by sea, however, and so this chapter begins by describing the caravans that once transported goods and people across the desert, and continues with an account of the world's most romantic overland trade route: the Silk Road that led from China to the eastern shores of the Mediterranean Sea.

CARAVANS AND OASES

A journey across a desert must be planned carefully. No one should attempt such a crossing alone. A vehicle breakdown, illness, or injury could be fatal in a remote region with a hostile climate. As if the environment were not harsh enough, traditionally there has always been a risk of attack from bandits.

People have always banded together for safety and crossed deserts in groups. In the Middle East, *kārwān* was the Persian word for such a group of travelers. The word has entered the English language as *caravan,* and the beast of burden that carried both goods and passengers was the camel. There are two species of Old World camels; the dromedary *(Camelus dromedarius),* which has a single hump, and the two-humped Bactrian camel *(C. bactrianus).* Camels, most probably Bactrian camels, were domesticated in what is now central Iran by about 2600 B.C.E.; in January 330 B.C.E. when Alexander the Great (356–323 B.C.E.) attacked Persepolis, the Persian capital city, he ordered 5,000 Bactrian camels to carry away the treasure his army captured. A bas-relief from the palace of Ashurbanipal at Nineveh, carved in about 645 B.C.E., shows dromedaries being used in warfare. Camels have been working for people for a very long time.

When caravan routes across the Sahara became established in the seventh and eighth centuries, dromedaries were the animals that made this possible. Camels are so well adapted to desert conditions that they came to be known as the "ships of the desert" (see the sidebar on page 54). Caravans also crossed the Gobi and Takla Masan Deserts in central Asia using Bactrian camels.

Although the desert appears featureless, caravans followed particular routes. One of the earliest, established in the third millennium B.C.E., ran from Nubia in the south to Egypt in the north, passing through the Al Kharga oasis, in the desert to the west of Karnak and Luxor. This was called the Darb-el-Arbain route; the caravans moving along it carried wheat, plants, and animals, as well as gold and ivory. The journey took 40 days, so in time it came to be called the Forty Days Road.

Another ancient route ran from Yemen to Gaza and Damascus—its caravans carried gold, silver, and frankincense—whereas caravans conveying salt used other routes. There were several salt routes, with the most important being those between the salt mines at Tauoudenni in northern Mali and Bilma in northeastern Niger, and Timbuktu in southern Mali. Another route ran from Lake Chad to Bilma and from there across the Fezzan region of Libya to Tripoli.

Pilgrims undertaking the hajj had to arrive in Mecca by the seventh day in the month of Dhū al-Hijjah, and so pilgrim caravans would depart from their different starting points on particular dates. Apart from that, caravans did not run to timetables. Fewer traveled in summer, because although the camel can endure harsh conditions

THE SHIP OF THE DESERT

A dromedary *(Camelus dromedarius)*, the single-humped or Arabian camel, has been known to go 17 days without drinking in the hot desert summer; in winter some camels do not drink at all. Camels have even been known to refuse water despite having drunk nothing for two months. Camels achieve this feat by using water very efficiently. Their kidneys extract and recycle more water from their food than do those of animals adapted to wetter climates, and camel urine is approximately twice as concentrated as human urine. Camels also sweat very little until their body temperature rises above about 105°F (40.5°C). On summer nights a camel allows its body temperature to fall as low as 93°F (34°C). This means that in the morning its body takes several hours to warm sufficiently to trigger sweating. In winter, however, and in summer if it has plenty of water, a camel's temperature varies by only about 4°F (2.2°C) between day and night. When a camel loses body fluid, it loses it from its tissues and not from its blood, so its blood volume remains constant while its tissues shrink, making the animal thinner. This allows it to tolerate a loss of water equal to 25 percent of its body weight. A human who loses an amount of water equivalent to about 12 percent of body weight will die very quickly. When it does drink, a camel can consume about one-third its body weight in water in the space of 10 minutes and a camel has been observed to drink 27 gallons (103 l) of water in that time.

The camel's hump is made mainly from fat. It is a food store that the animal metabolizes when food is not available.

When camels rest, they lie down with their legs folded beneath them, minimizing the amount of their bodies they expose to the heat. They lie side by side, shading each other, and they all face the Sun, which also minimizes the surface area they expose to the Sun. As the Sun moves across the sky, all the camels turn so they remain facing it. Dromedaries, but possibly not Bactrian camels, have pads of horny tissue on their chests and knees that insulate them from the hot sand when they lie down.

The camel's coat is thick. It traps a layer of air next to the animal's skin. The outermost hairs absorb the heat from the Sun so it does not penetrate to the skin. At the same time, when the camel sweats, its own body heat, rather than the sunshine, evaporates the perspiration, thus cooling the skin. Because the camel stores its body fat in its hump, it has no layer of fat beneath the skin that would prevent heat leaving its body.

Camels possess hooves, but they are at the tips of the digits and serve no function. Thick pads of skin cover the camel's toes and it walks on these pads. Its feet are very broad, making it less likely that they will sink in loose sand. Camels have long hairs protecting their ears from windblown sand, and their double layer of long eyelashes protects their eyes in the same way. When the air is full of dust and sand, a camel can close its nostrils.

they will weaken if they cannot eat for long periods and in extreme conditions they may perish. At other times of year, drivers would wait until they had enough cargo and passengers to make the trip worthwhile, and before a journey their camels spent several months grazing on good pasture to fatten them.

Passengers rode in panniers, which were large baskets or bags carried one on each side of the camel. A camel usually carried a load weighing about 350 pounds (160 kg). Caravans were large. Those in northern China and central Asia usually consisted of about 150 Bactrian camels, but caravans in the Sahara and Middle Eastern deserts usually had about 1,000 camels—sometimes as many as 12,000. Caravans are often depicted as strings of camels walking in single file. Chinese caravans were often like this, but the huge dromedary caravans were made up of strings of about 40 camels, linked together by ropes from the saddle to the nose ring of the camel behind. The strings walked three or four abreast.

Where possible, caravan routes called at oases. The Bilma–Tripoli route, for instance, had stopovers at the oases of Marzūq and Sabhā. Deserts are dry, but there is water beneath the surface that flows through *aquifers* in the permeable sand and gravel of the desert, above a layer of impermeable material. In some places natural depressions in the land are deep enough to be below the surface of that underground water. Water then lies at the surface. That is the most common type

A natural depression is deep enough for its bottom to lie below the surface of water flowing through an underground aquifer. Water then lies at or close to the ground surface, within the reach of plant roots.

of *oasis,* a word derived from two Coptic words, *oueh* ("to dwell") and *saa* meaning ("to drink"). Once water is at or close to the ground surface, plant roots are able to reach it, so plants flourish at oases and the land is often farmed. The illustration on page 55 shows an oasis of this kind. In Asia and in the Atacama Desert of South America there are oases fed by rivers flowing from the mountains.

THE CARAVANSERAI

A camel caravan would travel at a steady two to three MPH (3–5 km/h). Camels, mules, and horses loaded with goods or passengers were able to maintain this speed for hour after hour. In very hot weather the caravan would move by night and rest during the day. Eventually, however, the animals and their attendants and passengers needed to rest. At intervals along the caravan routes there were overnight stops, the equivalent of modern motels. The Arabs called them inns, for which the Arabic word is *kān,* or in English *khan.* In Turkish it is *han.* The Persian name for them was *kāwānsarāy,* transliterated into English as *caravanserai.* The word means a building with enclosed courtyards *(sarāy)* where caravans *(kāwān)* could rest.

Caravanserais were provided every 20–25 miles (32–40 km), which is the distance a caravan could cover in about 10–12 hours. Some caravanserais were a long way from the nearest town, but where they were associated with a town or village they were built outside the community's walls, simply because of their size. Most caravanserais had high square or rectangular walls enclosing a large open courtyard. The architecture was often elaborate and the walls and arches highly decorated. There was a single entrance that was large enough to allow laden camels, mules, and horses to enter. The establishment was open from dawn to dusk, at which time the door would be closed and secured by a thick chain.

The central courtyard was large enough to accommodate several hundred animals. Surrounding it there was a covered walkway, resembling a monastic cloister, with stalls opening off it where animals could be stabled and rooms where merchandise could be stored. Stairs led to an upper floor where there was a similar walkway with bedrooms for the travelers.

Caravanserais were owned by the nearest community and were open to everyone. They provided water for drinking, cooking, and washing. Some provided baths. Usually there was fodder for animals; the more elaborate establishments had restaurants, as well as shops where merchants could sell their wares to other travelers. More often, however, visitors provided their own food.

Many of the old caravanserais have fallen into ruin, but not all of them. Where the fabric of the buildings has survived, they have been converted into luxury hotels.

SILKS AND SPICES

Most buildings are constructed from materials such as wood, stone, or clay (in the form of bricks), all of which are locally plentiful. Only important public buildings and buildings owned by the very rich use materials that have been transported over long distances, and those materials are often displayed prominently, as decoration. That is because transporting bulky materials greatly increases their cost. The huge size of modern oil tankers and bulk carriers makes it economic to ship low-value commodities such as oil and grains across oceans, but this is a recent development. In the past, long-distance transport by caravan or wooden sailing ship was always expensive, and to cover the cost merchants sought the most valuable cargoes. Silks and spices were always high on the list of goods for which the supply could never match the demand, so their prices were always high.

Silken cloth is woven from threads taken from the cocoons of several species of moths, but originally and most importantly from the silk moth, *Bombyx mori,* which yields silk of the highest quality. Silk was first produced in China, where it was certainly being made by about 4000 B.C.E. and possibly as early as 5000 B.C.E., and the silk moth has been domesticated for so long that it is now incapable of surviving without human assistance.

Because of its fine texture and ability to absorb brightly colored dyes, when Europeans first saw silk they thought it a fabulous material and at one time it was literally worth its weight in gold. They had no idea how it was made. The Romans believed the Chinese made it from the leaves of certain trees, but the Roman writer Pliny the Elder (23–79 C.E.) knew that silk moths made the thread from which the cloth was woven. Neither Pliny nor anyone else in Europe had ever seen a silk moth or knew how they were cultivated, because the Chinese guarded the secret closely. The penalty for revealing the secret to a foreigner was death by torture. Despite this, the secret escaped. The Koreans were producing silk by about 200 B.C.E., and soon after 300 C.E. silk cultivation began in India. It was also in about 300 B.C.E. that a Japanese party captured four Chinese girls and a quantity of silk moth eggs. They forced the girls to explain how to tend the larvae—silkworms—and extract silk from the cocoons. That was the start of Japanese silk production, but silk manufacture increased greatly in the eighth and ninth centuries, when trade increased between Japan and China.

In China, for 1,000 years only members of the royal family and the highest officials were permitted to wear silk, but gradually the fabric became more abundant, until many people wore it. Then silk came to be used for other purposes, including the making of paper. By the time of the Han dynasty (202 B.C.E.–220 C.E.) silk was being used as a form of money. People were paying their bills with silk, and officials sometimes received their wages in this form.

China exported finished silk to the west, but in the sixth century wars between the Byzantine and Persian Empires made the trade hazardous and the Byzantine emperor Justinian I (527–565) sought alternative routes that would bypass Persian territory. A Christian group called Nestorians attracted supporters in the fifth century. They were denounced as heretics in Byzantium, but the Assyrian Christian Church accepted them, as did the Persians. In 552, Justinian sent two Nestorian monks to Central Asia to discover how silk was made. They obtained eggs, hid these eggs in their bamboo walking staffs, and returned to Constantinople. By the time they arrived the eggs had hatched but had not yet formed cocoons, and before long local weavers were producing silk fabrics for the emperor. Later, with the spread of Islam, Arabs brought silk production to the lands they occupied, and still later the method reached other parts of Europe. China had lost its monopoly, but it continued to export silks of the highest quality.

Oriental spices and herbs were also highly prized. They added flavor to the bland diet of northern Europe and disguised the bad tastes of food that had passed its prime but was, though disgusting, not yet poisonous. Spices were made into incense, and both herbs and spices perfumed stored fabrics and helped protect them against damage by moths.

The Egyptians imported spices from the Land of Punt (see "Egyptians on the Nile" on pages 2–4) and Arabia. These may have originated in India. The Greeks and Romans also imported Oriental spices, both nations trading directly with India and also importing frankincense, myrrh, and other materials for making incense from the Horn of Africa and southern Arabia.

These goods traveled along a network of routes known as the Incense Road. No individual merchant traveled the entire distance, however, and goods were bought and sold at trading posts along the way. The traders would never reveal the source of their commodities, for fear that the buyers would start dealing directly with the next merchant higher up the chain, bypassing the middleman. Consequently, legends based on travelers' tales peopled the spice-producing lands with fantastic beings.

THE SILK ROAD

The Xiongnu was a confederation of nomadic tribes that had conquered more than 20 nations in central Asia by about 180 B.C.E. When the Chinese attempted to block the supply of Chinese weapons to the Xiongnu and impose a trade embargo, the tribes invaded Chinese territory. In 198 B.C.E. the Chinese finally agreed to a traditional type of peace settlement with the Xiongnu, based on marriages between Chinese princesses or high-ranking courtiers and Xiongnu nobles, and the payment of tribute by China. The agreement was meant to end the conflict, but Xiongnu raids continued until, in 133 B.C.E., the emperor Han Wudi (156–87 B.C.E.) ordered the first in a series of attacks on the Xiongnu. These eventually defeated all of the Xiongnu and established Chinese control over the land as far west as Lop Nor (40°N 90°E) on the eastern edge of the Takla Makan Desert.

The emperor had earlier sent his general Zhang Qian on a diplomatic mission to forge alliances with nations to the west. The Xiongnu captured Zhang Qian and his caravan and held them prisoners for 10 years, but at last they escaped and he completed his journey before returning to the Chinese capital Ch'angan (modern Xi'an)—and being captured along the way by Tibetan tribes and held captive for a year. His mission made no new allies for the emperor, but Zhang brought back a great deal of information about the countries to the west and their inhabitants. This encouraged the emperor to send more missions that would establish political and commercial contacts. Goods began to move in both directions between China and the central Asian states and contacts still farther to the west allowed Asian goods to reach Europe.

Regular trading routes developed and in about 100 B.C.E. these became what is now called the Silk Road. It was the German geographer and traveler Ferdinand von Richthofen (1833–1905) who first called this route the "Silk Road," in his work *China: Ergebnisse eigener Reisen und darauf gegründeter Studien (China: The Results of My Travels and the Studies Based Thereon),* published in five volumes with an atlas between 1877 and 1912; he used the term because silk was the most valuable commodity that moved along the route and the reason why it was established, and silk was still being used as currency.

As the map on page 61 shows, there were two Silk Roads, one to the north and the other to the south. Both began at the capital Ch'angan (Xi'an), where smaller routes converged from the east, bringing goods destined for the long journey. The route led northward, meeting and then following the Great Wall past the Nan Shan Mountains. It skirted the southern edge of the Gobi Desert and passed through Dunhuang, a town built around an oasis in the Kansu-Sinkiang Desert. West of Dunhuang the Silk Road divided into two branches. One passed the northern edge of the lake of Lop Nor and skirted the Tarim Basin on the northern side. The Tarim Basin occupies the center of the Takla Makan Desert.

From the Takla Makan the route led to Kashgar, after which it climbed into the Pamir and Karakorum Mountains on the border between the Sinkiang Uighur Autonomous Region of China and Tadzhikistan. The route then led to Samarkand and Bukhara, in Uzbekistan. From there, one branch led northward to the Mongol city of New Serai, to the north of the Caspian Sea. (New Serai has since vanished, but it was the capital of the Golden Horde.) The other branch crossed to the south of the Caspian Sea through northern Iran skirting the Dashte Kavir Desert, crossed Iraq and Syria, passed through Damascus, and reached the Mediterranean coast at the port of Antioch, the city in southern Turkey that is now called Antakya. Passengers and goods were carried from there by sea to Italy. The southern route separated at Dunhuang and passed through Khotan (now Hotan), then the capital of the Buddhist kingdom of Khotan, on the southern side of the Tarim Basin. From there it went through Bactria, now part of Iran, to the oasis-city of Merv, near Mary in

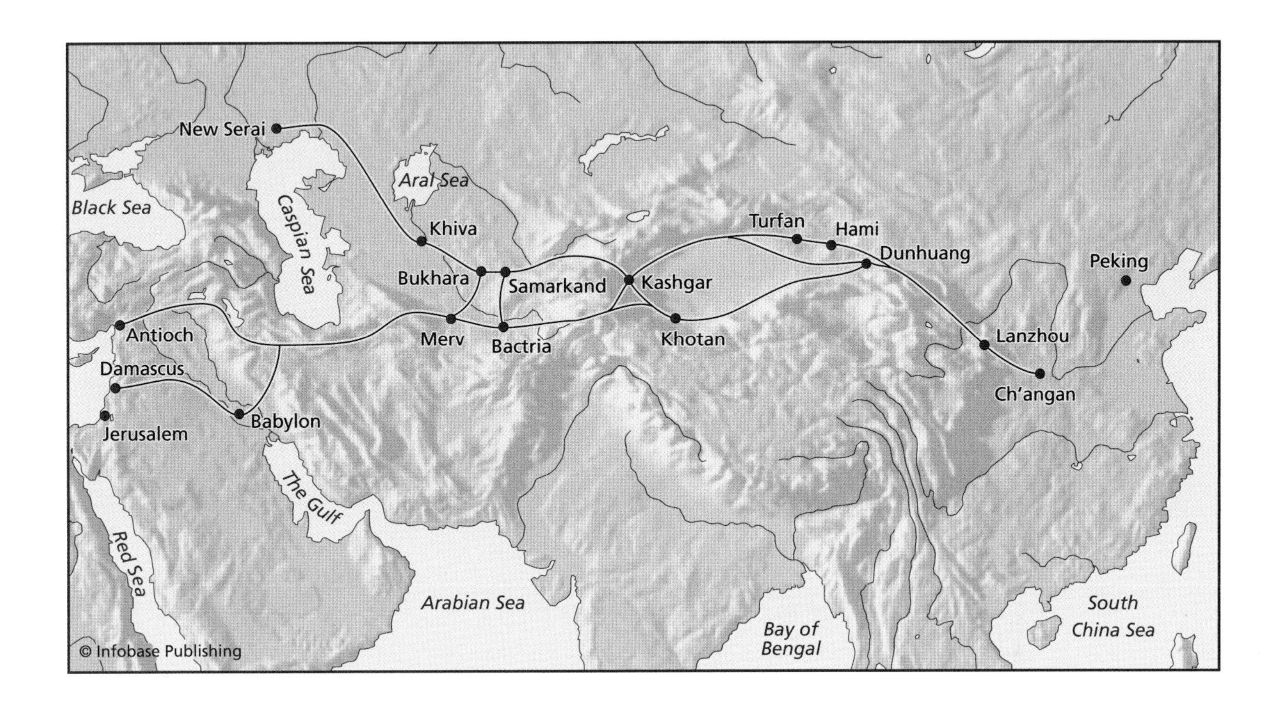

The Silk Road began at the city of Ch'angan (now called Xi'an) and followed two principal routes across Asia to the eastern shores of the Mediterranean, from where it continued by sea to the city-states of Italy.

modern Turkmenistan. For a time in the 12th century, Merv was said to be the world's largest city, and its ruins are now a World Heritage Site. Roads linked the northern and southern routes between Samarkand and Bactria and Bukhara and Merv.

With the decline of the Roman Empire and the rise of militant Islam, traveling the Silk Road became dangerous and the route fell into disuse, and sea routes became more important. People still traveled it during periods of political stability, however. When Marco Polo made his journey (see "Marco Polo and His Travels in Asia" on pages 123–124), the Mongol emperor Kublai Khan had established peace in the lands he crossed. There are now plans to restore the Silk Road as the Trans-Asian Highway.

FROM ASIA TO VENICE TO THE NETHERLANDS

By early in the 16th century Venice was one of the most important silk-producing states in Europe. In about 1500 the Venetian ruler, Doge (Duke) Leonardo Loredan (1436–1521), began to appear in public wearing robes made from gorgeous velvets and damasks, deliberately advertising and promoting the Venetian silk industry. These fabrics were already popular throughout Europe. They were manufactured in the Ottoman Empire (see the sidebar on page 63) and had long been imported by Venetian merchants, who then sold them to the rest of Europe. Venetian weavers had begun making them around 1450, and the doge wished the relatively new industry to prosper.

In those days Italy consisted of a number of independent city-states and Venice was one of the wealthiest and most powerful. It was one of the four maritime republics (the others were Genoa, Pisa, and Amalfi) that depended on overseas trade and sea power. Venice had a monopoly on sea salt in an age when salting was the principal method for preserving meat, fish, and some vegetables. Its main rival, Genoa, had a monopoly on alum, used as a *mordant*—a substance used to make colors fast—in dyeing.

Both Genoa and Venice had links with the Ottoman Empire. They maintained forts along the eastern Mediterranean coast to guard the entrances to trading ports that they controlled. The Turks began to occupy these colonies from 1451 during the reign of the sultan Mehmed II (1432–81), and his conquest of Constantinople in 1453 marked the end of the Byzantine Empire. Mehmed sought financial gain and the Genoese and Venetian merchants were prepared to pay large sums to the Turks to secure trading agreements that would allow their alum and salt imports to continue. The Genoese secured a solemn undertaking that allowed them to trade within the Ottoman Empire and to practice their Catholic faith. The agreement also granted them exemption from the Ottoman system of conscripting Christian boys for conversion to Islam and recruitment into the military as *janissaries* or into the civil service. (Janissaries were crack infantry units that formed the Ottoman sultan's bodyguards and household troops.) Early in 1454, Venice signed a treaty of peace and friendship under which the Turks undertook not to initiate or support attacks against Venetian citizens. This allowed the Venetians to maintain commercial establishments inside the Ottoman Empire, including Constantinople, where they retained a commercial agent in return for an annual payment of 200,000 gold ducats. The agreement required the Venetians to pay only 2 percent customs duty on their goods entering or leaving Ottoman territory.

Salt and alum were essential commodities, but they served an additional purpose: Because they were heavy and bulky, they could be

THE OTTOMAN EMPIRE

Early in the 14th century, Anatolia, in Turkey, comprised a number of small, independent Muslim states that had emerged as the Byzantine Empire weakened. Osman I (1259–1326), who ruled one of these states in western Anatolia, expanded the borders of his realm westward, established its capital in the city of Bursa, and founded what would become a Turkish empire that is usually known as the Ottoman Empire (the term derived from his name).

Osman died in 1326, and in the succeeding century the empire extended itself across the countries of the eastern Mediterranean and the Balkans. In 1387 the Turks captured Thessaloniki from the Venetians, in 1389 they captured Kosovo, and in 1453 they captured Constantinople (renamed Istanbul in 1930). They then moved into North Africa and the Ottoman navy fought the Italian states for supremacy in the Mediterranean, Aegean, Black, and Red Seas, and Portugal for control in the Indian Ocean. During the 16th century Ottoman forces conquered Hungary and much of the Middle East, including Persia (Iran) and Mesopotamia (Iraq). At the height of its power, in the 16th and 17th centuries, the Ottoman Empire covered most of southeastern Europe, the Middle East as far as the Caspian Sea and Persian Gulf, the lands on both sides of the Red Sea, Egypt, and North Africa as far west as Algeria. The Empire was ruled by a sultan, advised and supported by viziers, who headed the military class. Provinces and sub-provinces had governors, who were also members of the military class. Religious and civil leaders held their positions under warrant from the sultan.

An attempt to conquer Russia was defeated by the forces of Ivan the Terrible (1530–84) at the Battle of Molodi in 1572, and the Ottoman expansion into Europe was halted in 1683 at the Battle of Vienna. At the Battle of Lepanto in 1571, an alliance of Catholic European navies led by Don Juan of Austria (1547–78), the half brother of Philip II (1527–98) of Spain, defeated the Ottoman navy in the Mediterranean. Its expansion checked, the Ottoman Empire entered a period of stagnation that lasted from about 1700 to 1830. During this period Austria took control of much of the Balkans, parts of Egypt and Algeria fell under British and French control respectively, and there were repeated wars with Russia and uprisings in Serbia. Greece declared war on Turkey in 1821. There were also popular movements calling for modernization inside Turkey itself.

The decline of the Ottoman Empire continued through the 19th and early 20th centuries, as the area under Turkish control grew progressively smaller. There was war with Italy in 1911 and 1912, and with the Balkans in 1912 and 1913. Ottoman Turkey entered World War I on the side of Germany and Austria–Hungary. The Ottoman government collapsed in November 1918, when British and French forces occupied Constantinople, and at the end of the war what remained of the Ottoman Empire was partitioned into independent countries. These included the states of the modern Arab world and the Republic of Turkey. The League of Nations granted Britain and France mandates over Syria, Lebanon, Mesopotamia, and Palestine.

used to ballast ships that were carrying silks, spices, and gemstones from the Middle East to Europe. These three cargoes were much more valuable than salt and alum, but they were light, and without the ballast the Italian ships would not have been seaworthy.

Trade was of the greatest importance to Turkey as well as to Venice, and silks and spices moved in both directions. Merchants established a commercial world in which Jews, Christians, and Muslims lived and worked side by side, and all of them grew wealthy. As silk had in China, spices became a kind of currency in Europe. In the 1530s Flemish tapestry weavers agreed to a commission from the king of Portugal on the condition that the work be paid for in pepper. Sugar was equally valuable. The queen of Portugal made charitable donations to convents in the form of sugar, which the nuns sold in order to provide themselves with items they needed.

Venice was the leading European sea power in the Mediterranean. Flanders was the most important commercial power in northern Europe, but during the 16th century it was overtaken by the Netherlands. All of the Low Countries—part of modern Belgium, and all of the Netherlands and Luxembourg—had fallen under Spanish rule as a result of inheritance and conquest, and the Spanish united them into a single state. They rebelled, and the Eighty Years' War against Spanish rule lasted from 1568 to 1648. During that time Dutch merchants began trading all over the world. In 1600 three Dutch ships sailed to the islands that became the Dutch East Indies (now Indonesia). The Dutch traded for spices, and in 1602 they founded the VOC (Vereenigde Oost-Indische Compagnie, or United East India Company), which had a monopoly on Asian trade. They also imported spices from India and traded in southern Africa and in the West Indies; in 1621 they established the GWIC (Geoctroyeerde Westindische Compagnie, or Dutch West India Company), which had a trade monopoly similar to the one held by the VOC. They conquered Portuguese settlements in Sri Lanka, the East Indies, Brazil, and Angola. They also established Nieuw-Nederland (New Netherland), a colony that existed from 1609 until 1667 in North America, with a town located outside Fort Amsterdam on Manhattan Island that they called Nieuw Amsterdam. In 1664 it came under British control and the British renamed the city New York.

By the middle of the 17th century there were Dutch trading stations throughout the world and the Netherlands was Europe's leading maritime and commercial power. This led to conflict with the English, whose trading empire was also expanding, and two wars between the Dutch and English were fought from 1652–54 and 1664–67. A third Anglo-Dutch war, from 1672–74, was part of a larger war between the Netherlands and France. A fourth Anglo-Dutch war, from 1780–84, followed Dutch recognition of the United States as an independent nation. This final war greatly weakened the Dutch economy and marked the end of the nation's commercial supremacy.

CRETANS AND PHOENICIANS

Europe's earliest civilization developed on the island of Crete in the eastern Mediterranean. The earliest evidence of occupation is about 7000 B.C.E., but Cretan civilization began in about 2600 B.C.E., at the beginning of the Bronze Age, with the rise of the Minoan culture, named after its legendary ruler Minos. Minos was the son of Zeus, ruler of the gods, and of Europa, a Phoenician princess, and he gained the Cretan throne with the help of Poseidon, the god of the sea. *Minos* was probably a title borne by all of the priest-rulers of the palace of Knossos, which was at the center of the civilization and culture. After his death, the legendary Minos entered Hades, where he became a judge of the dead.

From its start, the Minoan civilization was a maritime power, and under Minos it asserted its control over the islands of the Aegean, establishing Minoan colonies on several of them, the most famous of which was on the island of Thera (also called Santorini). Minoan ships were built from cedar wood and used sails as well as oars. They had a high prow, often ending in a two-pronged fork. Some had a high stern, while in others the stern sloped downward to sea level. There are also pictures of Minoan ships with devices in the shape of fishes dangling from the prongs of the prow. These may have indicated the wind direction. Pictures from Thera show ships with a flat feature projecting from the stern just above the waterline and disk-shaped objects attached to the prow. It is possible that a strong wind from the side of the vessel would push against the objects on the prow, while the stern feature would create drag, the combined effect turning the ship until its stern pointed into the wind. There were usually two steering oars, and an awning that made a cabin for passengers.

Depictions of Minoan ships suggest that they were used for trading. They carry no visible armament or armed troops, and there is no ram, although they must have been able to defend themselves against pirates. Minoan power appears to have been economic and great enough to allow the Minoan people to live in coastal cities with no protective walls.

The Minoans imported Egyptian goods and copied them, and exported their own distinctive pottery to Egypt, Cyprus, and Syria. Minoan products were known as far away as the Euphrates, and Cretan metalworkers made daggers, in a style derived from Syria, and exported them to Cyprus, possibly in exchange for copper, although there were copper ores on Crete. The Minoans also imported tin, which was an important commodity in the days when bronze—an alloy of tin and copper—was in widespread use. They also grew and traded in saffron, which was in demand as a dye as well as a flavoring.

The Minoan civilization ended abruptly at some time between 1627 B.C.E. and 1600 B.C.E. when a volcano erupted on the island of Thera. This was one of the largest eruptions in history, hurling 14 cubic miles (60 km^3) of rock, dust, and ash into the air. The Thera settlement was buried in pumice. Crete was not seriously affected by the eruption itself, but geologists believe that a large earthquake that preceded the eruption devastated the island.

Despite the disappearance of the old civilization Cretan society recovered and Cretan ships continued to trade throughout the Mediterranean. A hemispherical bronze bowl that was made in about 900 B.C.E. and found near the site of Knossos bears Phoenician writing. The oldest inscription written in Greek that has ever been found is written on a jug that originated in Crete. Phoenician pottery has also been found in southern Crete, and there is evidence that the Phoenicians had a colony on the island. Contact between the two cultures began much earlier than that, of course, because Europa, the wife of Minos, was Phoenician. Some scholars believe that it was the Minoans who taught shipbuilding and seamanship to the Phoenicians and the Greeks.

TIN FROM CORNWALL, IVORY AND PEACOCKS FROM ASIA

In about 1100 B.C.E. the Phoenicians established a colony at the place they called Gades (modern Cádiz). Phoenician merchants were keen to trade in any commodity likely to make them a profit, and Phoenician sea captains were skilled and brave. They ventured into the Atlantic, crossed the treacherous Bay of Biscay, and reaped a rich reward. They discovered a long peninsula where people were producing one of the world's most important metals: tin. The Phoenicians had reached Cornwall, in the southwest of mainland Britain.

The best quality bronze is an alloy of copper and tin. An inferior bronze is made from arsenic and copper, but tin is the preferred alloying metal. In the Bronze Age, therefore, tin and copper were essential. Copper was fairly abundant and Cyprus was the principal source for the Mediterranean nations. Tin is much less common. It does not occur naturally as the pure metal, and the mineral cassiterite (SnO_2) is the only important ore. Cassiterite is highly resistant to chemical attack and it is totally insoluble. Consequently, it tends to survive when the other minerals around it have dissolved and been washed away and forms *alluvial deposits*—materials transported by streams and rivers and found on riverbeds and floodplains. Such deposits occurred in only a few places, however, and Cornwall was one of them. In about 2300 B.C.E., people in various parts of Britain, including Cornwall, began making a distinctive type of pottery that has given them the name Beaker People. The Beaker People were employing new technologies, and they were the first people in Britain to use bronze. At first they obtained cassiterite by streaming it, and perhaps the hard, heavy mineral was a novelty stone that they found while searching streams for their principal goal—alluvial gold. In time the alluvial deposits were depleted and the tin producers began mining cassiterite. By the time the Phoenicians arrived, the Cornish people had been smelting cassiterite to obtain tin for a very long time.

The links between Cornwall and Phoenicia are firmly established in local Cornish tradition, but although the link may well have existed there is very little archaeological evidence for it. That is probably because the Phoenicians kept the source of their tin a closely guarded secret. They said it came from the Cassiterides or "tin islands," a name that suggests a group of islands off the coast of northwestern Europe. This was deliberate misinformation, for there are no islands with cassiterite deposits off the European coast: Though they are only about 30 miles (50 km) from the westernmost tip of Cornwall, the Isles of Scilly have no tin ore.

The Greek philosopher and geographer Hecateus of Miletus (ca. 550–ca. 476 B.C.E.) claimed to know of northern islands where tin was produced, although he does not depict them on the map of the world that he drew in about 500 B.C.E. The illustration below shows a 19th-century reconstruction of that map. It is interesting because it depicts the Mediterranean, Red, and Black Seas fairly accurately and indicates that Hecateus knew of the existence and approximate location of the Caspian Sea, but it becomes very vague in its information about northern Europe, and it does not show the British Isles at all.

The Greeks established colonies along the coast of southern Europe and imported their tin from alternative deposits of cassiterite in northern Spain that were more easily accessible. Spanish tin was exported from the port of Tartessos in what is now Andalucía, but in the sixth century B.C.E. all records of Tartessos cease. It is likely that the Carthaginians from Gades had destroyed the city. The Carthaginians also controlled the Pillars of Hercules (Straits of Gibraltar) so the destruction of Tartessos gave the Phoenicians based at Carthage a monopoly of the tin trade. The Greeks were keen to break that monopoly and in about 325 B.C.E. the explorer Pytheas (his dates of

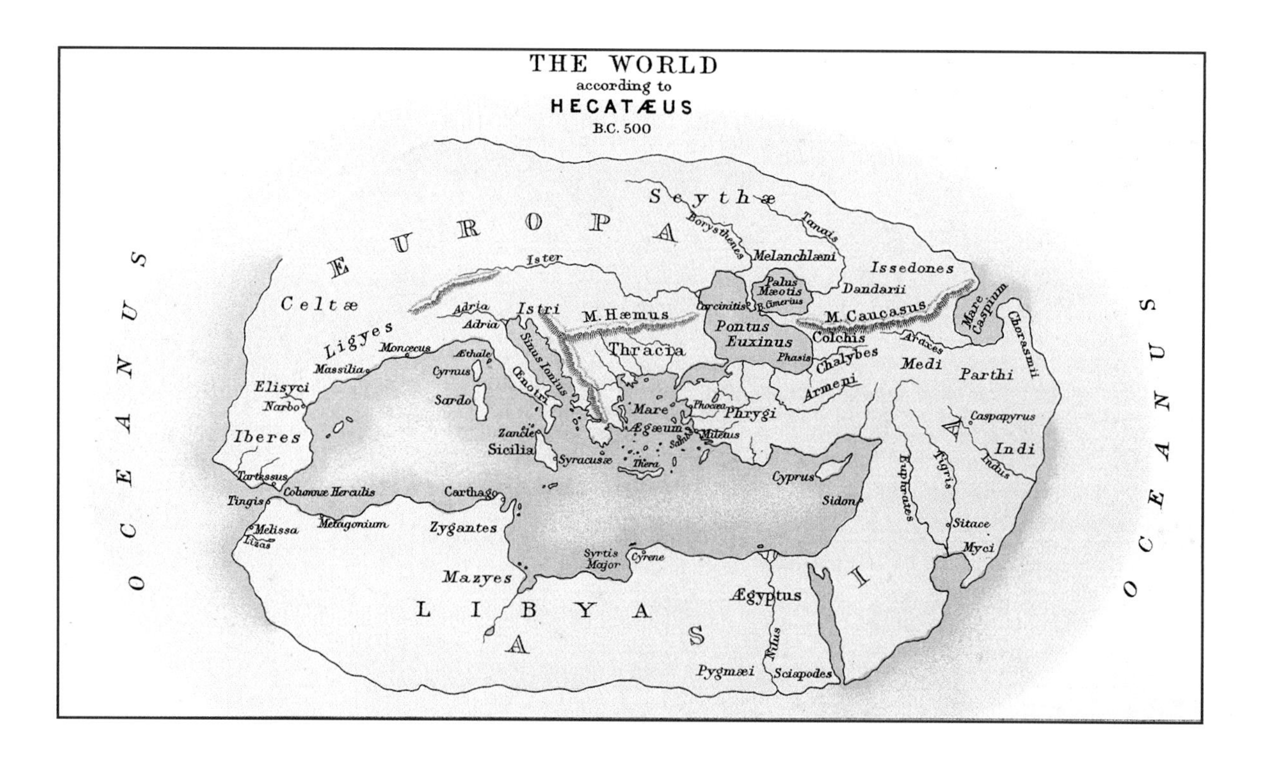

A 19th-century reconstruction of the map of the world drawn in about 500 B.C.E. by the Greek philosopher and geographer Hecateus of Miletus (ca. 550–ca. 476 B.C.E.) *(Granger Collection)*

birth and death are not known) set out from the Greek colony at Massalia (Marseilles) to seek a way to bypass the Carthaginians. He may have been able to pass through the Pillars of Hercules at a time when the Carthaginians were distracted by a war they were fighting with the city-state of Syracuse, in Sicily, or he may have traveled overland to the English Channel. His description of the journey is lost, and what is known about him today comes mainly from later writers. Pytheas was all too ready to believe improbable tales he heard from other travelers and his estimates of distances were wildly inaccurate, but it does appear that he visited Britain and that he was probably the first visitor to call it Britannia. He also described the tin miners. According to the first-century Greek historian Diodorus Siculus (whose dates of birth and death are also unknown), Pytheas describes the tin miners in the following words:

> The inhabitants of that part of Britain which is called Belerion are very fond of strangers and from their intercourse with foreign merchants are civilized in their manner of life. They prepare the tin, working very carefully the earth in which it is produced. The ground is rocky but it contains earthy veins, the produce of which is ground down, smelted and purified. They beat the metal into masses shaped like knuckle bones and carry it off to a certain island off Britain called Ictis. During the ebb of the tide the intervening space is left dry and they carry over to the island the tin in abundance in their wagons.

Belerion is Land's End, the westernmost tip of Cornwall. Ictis is St. Michael's Mount, an island off the south coast of Cornwall that is linked to the mainland by a causeway that is above water at low tide. The tin ingots were shipped across the Channel and then transported overland, mainly along river valleys, to the Mediterranean coast.

Ivory was always in demand and, therefore, always valuable, and the Egyptians were importing African ivory more than 4,000 years ago. At one time there were African elephants *(Loxodonta africana)* living in northern Syria. Their tusks supplied the material for a highly profitable ivory trade controlled by the kings of Aleppo, a place strategically placed to exploit markets in both the Mediterranean and Mesopotamia. The Syrian elephants were hunted for ivory so relentlessly that they were extinct by 500 B.C.E.

The Harappan civilization flourished between about 2500 B.C.E. and 1700 B.C.E. in the Indus Valley, on the border between Pakistan and India, centered on the two cities of Harappa and Mohenjo-Daro. The Harappan people traded by sea with Egypt, Minoan Crete, southern Arabia, and with the Mesopotamian kingdoms in what is now Iraq. Their ships sailed to Bahrain, from where goods were transported inland and there were probably Harappan trading posts in Mesopotamia. Harappan seals and other items from that time have been found throughout the Middle East and pictures of ships appear on seals found at Mohenjo-Daro.

The Harappan traders had much to offer. They exported copper, ivory, and wild animals to stock royal menageries, and among the animals they sent westward was that most spectacular cousin of the pheasants, the peafowl *(Pavo cristatus)*. Today the male peacock is India's national bird, and it has always been highly prized, with peacocks being presented to kings as gifts, together with gold and silver. King Solomon (1020–980 B.C.E.) received them. In the Old Testament, the book of Kings records that:

> For the king had at sea a navy of Tharshish with the navy of Hiram: once in three years came the navy of Tharshish, bringing gold, and silver, ivory, and apes, and peacocks. So king Solomon exceeded all the kings of the earth for riches and for wisdom. (*1 Kings* 10: 22–23)

The navies of Tharshish and Hiram refer to Phoenician vessels. The Phoenicians also brought peacocks to Tharsis, a town in southern Spain. By the fifth century B.C.E. the descendants of peacocks imported from Asia Minor (Turkey) were a common sight in Athens, and Alexander the Great (356–323 B.C.E.) brought these birds from India to Europe. In the first century C.E. the Romans began breeding peacocks for their meat, but they also sacrificed them to Juno, the goddess of matrimony and motherhood.

Throughout history, sea trade, based on luxury items as well as necessities, has linked communities across the world. At the same time, sailors and merchants carried information about themselves, their homelands, and cultures, and returned with stories about the peoples and lands they had visited. Traders were also explorers.

DISCOVERY OF THE MONSOON

In winter, cold air subsides over central Asia, producing high atmospheric pressure near the surface. Air flows outward and sinks down the southern side of the Himalayan Mountains to produce northeasterly winds over India. These winds bring very dry air and produce the season known as the dry, winter, or northeasterly monsoon. In summer, the distribution of pressure reverses. Warm air rising over central Asia produces low surface pressure and air moves into it from the region of higher pressure over the Indian Ocean. This brings southwesterly winds to India, and because they have crossed the warm ocean these winds produce heavy rain as they rise to cross the land. This is the wet, summer, or southwesterly monsoon. The illustration on page 72 shows the pressure and wind patterns of the two monsoon seasons. The intertropical convergence zone is where the northeasterly and southwesterly winds meet.

Ancient seafarers cared nothing for the seasonal climate of India, but they were keenly interested in the prevailing winds. The discovery of the summer monsoon winds was of major importance, because it informed merchants that summer was the best time to cross from Arabia to India in search of spices, and winter was the best time to return. The winds could not have been arranged more conveniently.

The discovery of the significance of the monsoon winds is usually attributed to a Greek navigator called Hippalus, who lived in the first or second century B.C.E. Hippalus was not the first Greek to sail to India and he did not discover the monsoon, which had been known for centuries. The Greek philosopher and geographer Poseidonius (ca. 135–ca. 51 B.C.E.) stated that the first Greek navigator to sail the monsoon winds across the Indian Ocean was Eudoxus of Cyzicus, who lived in the second century B.C.E. Poseidonius recounted that a shipwrecked Indian sailor had been rescued on the Red Sea coast and taken to Ptolemy VIII (ca. 182–116 B.C.E.), the king of Egypt. The sailor said he could guide a Greek ship to India, and Ptolemy appointed Eudoxus to command the vessel. Eudoxus made two voyages in 118 B.C.E. and 116 B.C.E., the second time without a guide, and Hippalus may have sailed with him.

What Hippalus discovered was how to exploit the summer monsoon winds; although he was the first Greek to do so, Arab and Indian navigators had already been using the winds for some time, as the voyages of Eudoxus demonstrate. His discovery is described in the *Periplus Maris Erythraei* (Periplus of the Erythraean Sea), a work written in the first century C.E. A *periplus* is a set of written instruc-

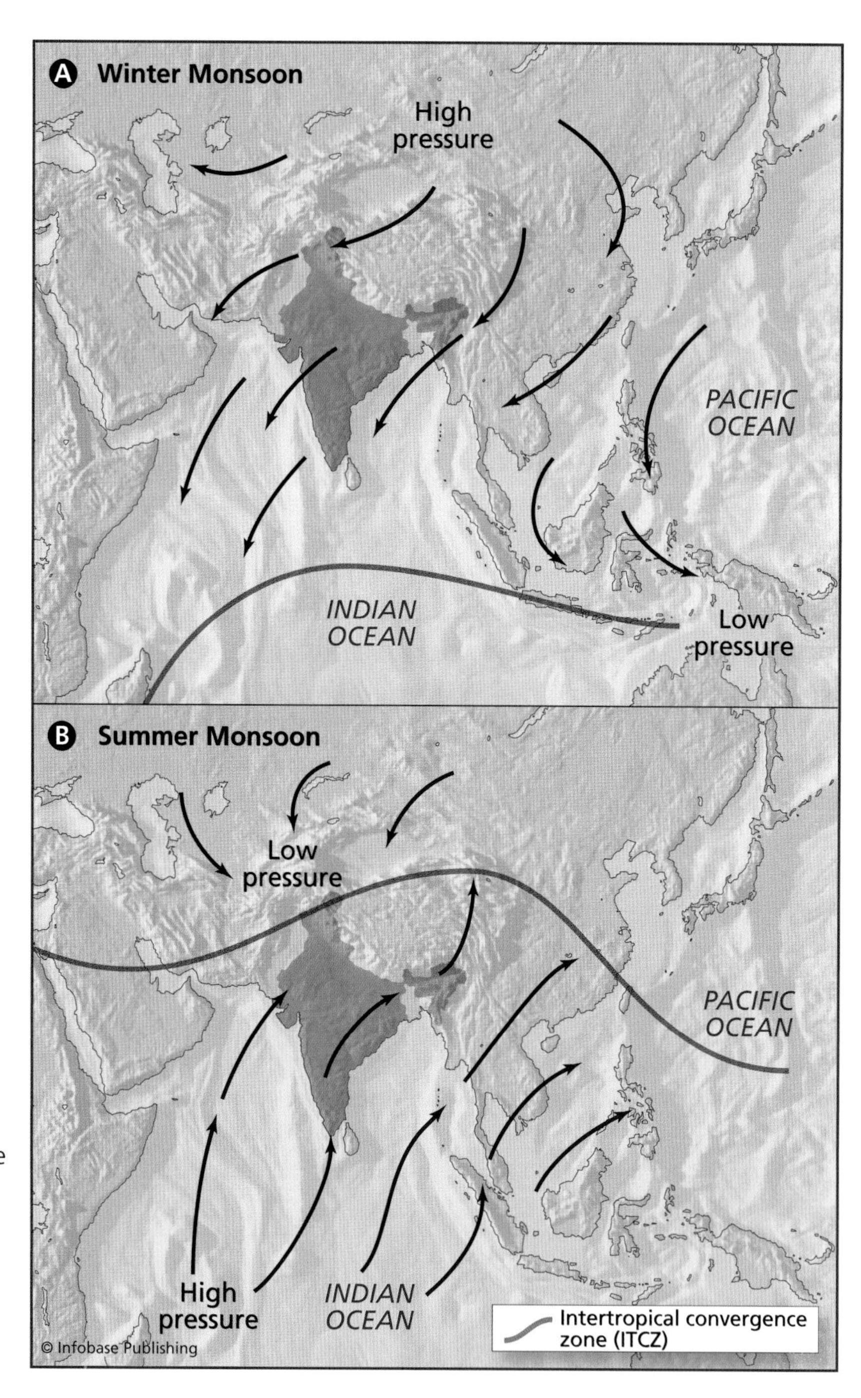

In winter, high pressure over central Asia produces subsiding air that spills down the southern side of the Himalayas and crosses India as dry, northeasterly winds. This is the dry, winter, or northeasterly, monsoon. In summer the pressure distribution reverses. Pressure is high over the Indian Ocean and low over central Asia, and southwesterly winds bring heavy rain to India. This is the wet, summer, or southwesterly, monsoon.

tions for navigators, literally a "sailing around." The Erythraean Sea was the Greek name for the combined Red Sea, Indian Ocean, Arabian Sea, and Persian Gulf. Hippalus may also have been the first Greek navigator to realize that the Indian coast runs from north to south and not from east to west. The *Periplus* describes his discovery in the following words:

> This whole voyage as above described . . . they used to make in small vessels, sailing close around the shores of the gulfs, and Hippalus was the pilot who by observing the location of the ports and the conditions of the sea, first discovered how to lay his course straight across the ocean. For at the same time when with us the Etesian winds are blowing, on the shores of India the wind sets in from the ocean, and this southwest wind is called Hippalus.

The Etesian winds are northerly winds that blow over the eastern Mediterranean between May and October. Hippalus had found that if a ship sets sail from Egypt in July it can head directly out to sea and the winds will carry it to India, and that this is much faster than the traditional route following the coast all the way. The *Periplus* also mentioned that the crossing was dangerous—the summer monsoon winds are strong and gusty, making the sea rough. Hippalus did not mention the return voyage using the winter monsoon winds. His discovery greatly boosted trade between Roman Egypt and India.

SEA ROVERS, PIRATES, AND PRIVATEERS

Phoenician seafarers were engaged in trade, not war, but they were not above stealing from their competitors. For as long as valuable cargoes have been carried by sea there have been pirates lying in wait to seize them. The Greek poet Homer referred to pirates in the *Odyssey,* where he called them "sea rovers."

Piracy was common when Rome ruled the known world. At one time warships from the Seleucid empire in Syria had protected merchant shipping, but the empire went into decline from about 150 B.C.E. and pirates based in the Balearic Islands, Crete, and Cilicia in southern Turkey went unopposed. The Romans had a powerful navy and could have checked them, but the pirates were useful because they supplied Roman landowners with slaves. Most of these were the crews of the merchant ships they captured, preferring the big, slow ships that carried Egyptian grain to Italy. Ordinary slaves were sold on the Aegean island of Delos, but wealthy captives were held for ransom.

In 75 B.C.E. Cilician pirates captured Julius Caesar (100–44 B.C.E.) and that is when they met their match. According to the biography of Caesar written by the Greek author Plutarch (46–ca. 122 C.E.), Caesar laughed when his captors demanded 75 talents as a ransom, and offered to pay them 50. He sent members of his party to nearby cities to raise the money, retaining two servants and one friend for company. He appears to have been perfectly relaxed and in control, ordering the pirates to keep quiet when he wanted to sleep and joining in their games. He wrote stories and poems that he read to them and sometimes he threatened to have them all hanged. They seem to have enjoyed what they took to be his childishly innocent behavior. In due course the ransom arrived and was paid, and Caesar was set free. Immediately he commandeered a number of ships and sailed in pursuit of the pirates, finding them still at anchor off the island where he had been held. He captured almost all of them, imprisoned them, seized all of their property, and asked the Roman governor to punish them. When the governor hesitated Caesar had the pirates removed from prison and he crucified all of them.

Although there have always been pirates—and are still—there was a "golden age" of piracy. It was very brief, lasting only from about 1680 until 1720, but it is the period in which the most famous pirate stories are set. The pirates wrote some tales of their exploits themselves. A small group of former pirates who had made their fortunes and accepted pardons lived on the edge of fashionable London society. More stories have been taken from *A General History of the Robberies and Murders of the Most Notorious Pyrates,* a book published in London in 1724 and written by "Captain Charles Johnson"; the work is sometimes called *Johnson's Pirates.* However, there is no record of any captain of that name during the early 18th century and the genuine pirates who told pirate stories were all dead before some of the true events recounted in the book. The suspicion is that Captain Johnson was really the journalist, novelist, pamphleteer, and spy Daniel Foe, who changed his name to Defoe (1660–1731). *Treasure Island* by Robert Louis Stevenson (1850–94), the most famous of all pirate adventure stories, uses material adapted from Defoe's work. It was first published in 1885 and immediately made Stevenson a literary star.

There were several reasons why seamen turned to piracy. The first was the desire of all governments to save money. Until the second half of the 19th century, navies recruited sailors in times of war and laid them off when their service was no longer required. This produced a peacetime pool of unemployed men who needed to make a living and who might agree to a venture that paid no wages but promised a share in the spoils.

Governments also hired privately owned and manned warships. The owner would be issued with a letter of marque, authorizing the use of the ship to harass the enemy in time of war or to carry out reprisals on ships of a specified nationality for offences committed by their governments in time of peace. The ship owner was designated a privateer; he received no payment but was entitled to a share in the prize money from the sale of captured ships and their contents. It was not until 1856, following the end of the Crimean War, that the Declaration of Paris brought an end to privateering. Even then, the officers and crews of naval ships continued to receive a share of the prize money from the auction of seized enemy vessels. The United States abolished that custom in 1904 and other governments abandoned the practice at around the same time.

Living and working conditions at sea were often appalling and the regimes brutal. When life onboard became unendurable crews sometimes mutinied, overthrowing their officers and seizing control of the ship. This was a capital offense, and to save their lives the mutineers might choose to take their chances as pirates. When pirates seized a ship that they intended to use, they would need a crew to man it. They would invite the existing crew to join them voluntarily, but those in key positions might be forced to become pirates on pain of torture or even death. Compulsion was always a valid defense against the charge of piracy.

The most successful of all pirates from that period was probably Bartholomew Roberts (1682–1722), known as Black Bart, who sailed the Caribbean and Atlantic from 1719 to 1722. He was born in South Wales and served as third mate on the *Princess,* a slave ship that was captured off the West African coast in June 1719 by the pirate Howell Davis; forced into piracy, Roberts quickly realized that he was suited to the life. When Davis was killed later the same month, Roberts replaced him as captain. He went on to command the pirate ships *Royal Rover, Royal Fortune I, Royal Fortune II,* and *Royal Fortune III*—all names that he gave to ships he had seized. By the time he was killed in a battle with HMS *Swallow* in February 1722, Roberts is alleged to have attacked and robbed approximately 400 ships.

BLACKBEARD AND CAPTAIN KIDD

Roberts was not the most famous of all pirates, however. Two men, Blackbeard and Captain Kidd, must share that honor—if *honor* is the appropriate word. Kidd was a privateer rather than a pirate, and so a respectable man. The illustration on page 77 shows him welcoming ladies on board his ship in New York, prior to embarking on the mission that would prove his undoing.

Blackbeard and Captain Kidd were two of the most notorious pirates to sail during the brief golden age of piracy. Blackbeard's real name was Edward Teach, or Thatch, or Drummond. He was born in about 1680, probably in Bristol, England, and went to sea at an early age. He may have sailed as a privateer in the Spanish West Indies and on the *Spanish Main,* which was the coast of Spanish America from Florida all the way to the mouth of the Orinoco River. Teach lived at New Providence, Bahamas, which was a center for pirates, and in 1717 he turned to piracy by joining the crew of Benjamin Hornigold (died 1719). When Hornigold accepted a pardon and retired, Teach took command of one of his prizes, a French slave ship called *La Concorde.* He armed the ship with 40 guns and renamed her *Queen Anne's Revenge.*

There is no evidence that Teach ever killed anyone who was not trying to kill him. He and his crew would board ships and seize all their valuables, weapons, and stores. If the victims resisted Teach would usually put them ashore in a remote spot and either burn their ship or commandeer it for his own use. All together, he may have robbed more than 45 ships.

Teach deliberately cultivated his savage image, and the nickname Blackbeard was his own invention. He grew a beard that covered much of his face and grew his hair long, braiding both beard and hair into pigtails with colored ribbons in the ends. Legend has it that before going into battle he placed pieces of smoldering hemp under the rim of his three-cornered hat, but this may not be true. He added to his ferocious appearance by wearing a crimson coat and having

Captain William Kidd (ca. 1645–1701) welcoming ladies aboard his ship, from Jean Leone Gerome Ferris's painting *Captain William Kidd in New York Harbour, 1696* (Hulton Archive/Getty Images)

two swords, several knives, and sometimes as many as six pistols tucked into his belt. His aim was to terrify his victims into surrendering before anyone was hurt. He was evidently popular with ladies and was believed to have had 14 wives.

In 1718 his ships blockaded the harbor at Charleston, South Carolina, capturing many prominent citizens from ships trying to enter the harbor and holding them hostage, demanding medical supplies. When these arrived he released all the captives. He then applied for the pardon that was offered to those who undertook to abandon piracy and settled down in Bath Town. Before long he returned to piracy, however, and Lieutenant Robert Maynard of the Royal Navy killed him in battle on November 22, 1718.

Captain William Kidd was born in about 1645 in Dundee, Scotland. He went to sea and rose to command his own ship, the *Antigua.*

In the 1680s he settled in New York and married Sarah Bradley Cox Oort. When Britain and France were at war in the 1690s Kidd became a privateer, commanding the *Blessed William,* a ship he and the crew had seized in 1689 by mutinying, and protecting English and American trade routes in the West Indies.

In May 1695, while in London seeking a new privateering commission, Kidd agreed to a deal proposed by Richard, earl of Bellomont, the newly appointed governor of New York and Massachusetts, and a number of his powerful friends including members of the British government. Kidd took command of the 34-gun *Adventure* galley, with a letter of marque to hunt down pirates in the eastern seas, with the understanding that he would seize their vessels and valuables. He sailed from London to New York to recruit more crew and on September 6, with a crew of 150, *Adventure* sailed from New York.

Kidd managed to capture only two vessels, the second and more important being the *Quedah Merchant,* carrying a rich cargo that included 70 chests of opium. Refusing a ransom from the ship's Armenian owners, Kidd decided to take the prize to St. Mary's Isle, Madagascar, an infamous pirates' lair, where most of his crew deserted to join the pirate Robert Colliford. The *Adventure* was leaking and Kidd, with the 13 men still loyal to him, barricaded himself on board the *Quedah Merchant,* leaving the *Adventure* to sink. Then, in order to find a way to get home, Kidd agreed to provide Colliford with weapons, ammunition, and money in return for Colliford fitting out the *Quedah Merchant* for an ocean crossing. News of the deal reached agents of the East India Company in Bombay and officials in London, and Kidd was marked down as a pirate. Kidd reached Hispaniola, where he obtained a small sloop and returned to New York, convinced that his powerful promoter would protect him. Bellomont listened to Kidd's story, but distanced himself from what was becoming a dangerous scandal and had Kidd arrested.

Kidd spent nearly a year in solitary confinement before he arrived in London on April 11, 1700, then spent a second year in solitary confinement in Newgate Prison, during which time he was denied clothing, writing materials, and legal advice. In March 1701 he was brought before the full House of Commons and urged to implicate his partners. Kidd maintained that since he had done nothing wrong, neither had they. Had he named them he might have been pardoned. As it was he was sent for trial at the High Court of Admiralty on

May 7, 1701. It was not until the night before the trial that lawyers appointed to defend him visited him in his cell, and told him he faced six charges on which he would have to defend himself. The first of these was murder, relating to an incident when after a series of confrontations with a mutinous gunner called William Moore, Kidd had thrown a bucket with iron hoops at him, fracturing Moore's skull. At the trial, Kidd was not permitted to call witnesses, he fumbled, and documents proving that he had seized the *Quedah Merchant* lawfully had mysteriously disappeared. The jury took only one hour to convict him and on May 23 he was taken, drunk and protesting his innocence, to be publicly hanged. His body was then gibbetted—hung in chains—at Tilbury Point on the Thames, as a warning to others.

The Art of Navigation

The very earliest sailors traveled along rivers, keeping the banks always in sight, crossed lakes, or ventured a short distance out to sea—but never moved out of sight of land. As they became more adventurous, they needed to find their way across open waters, where there were no familiar landmarks to guide them. This chapter explains some of the problems navigators had to solve and the way they did so. The chapter begins with a key discovery—of the Pole Star, Polaris—and goes on to explain the meaning of north and south and describes the invention of the magnetic compass. Obviously, a navigator needs to know the direction in which the ship is sailing, but that information is not sufficient in itself to ensure a safe landfall. The seafarer must also know the position of the ship and its speed. These called for different instruments, and the chapter tells of their invention and how they were used.

DISCOVERY OF THE POLE STAR

Very often, when a large ship enters unfamiliar coastal waters it takes on a pilot. Pilots are local navigators with an intimate knowledge of the approaches to a harbor. They must know the location of deep-water channels and hazards such as reefs and sandbanks, but basic piloting is the most ancient of all navigational techniques. It involves guiding the ship with reference to landmarks on shore.

Piloting allowed vessels to cover long distances, but thousands of years ago seafarers realized that the shortest voyage between two ports was seldom the one that followed the coastline. They had to take courage and sail over the horizon, out of sight of land. There were clues that these ancient navigators could use. In the Mediterranean, for instance, warm winds always blow from the south and cold winds from the north, so the wind direction often conveyed useful information, and it was information that could be extended. It may have been the Etruscans who invented the *wind rose,* which at first was a simple diagram showing the directions of the principal winds. The Etruscans lived in Italy and Corsica prior to the founding of Rome and they were sailing the Mediterranean by about 500 B.C.E. The illustration below right shows a modern wind rose diagram depicting the prevailing winds at a particular place. The diagram consists of a circle with straight lines radiating from the center. Each line represents a wind direction, and the length of each line indicates the frequency with which the wind blew from that direction over a reference period, usually of one month or one year. The sea surface also held information. Water currents followed known tracks that were often visible as changes in the color of the water.

Navigators also had astronomical clues. Some were obvious, but no less important for that. For example, the Sun rises in the east and sets in the west, and in the Northern Hemisphere it is due south at noon, so at noon the shadow of the ship's mast points north and south. The early navigators lacked clocks to tell them when it was noon, but they were able to observe when the Sun reached its maximum height above the horizon, which it does at noon. At night the stars also rise in the east and set in the west. The Greek historian

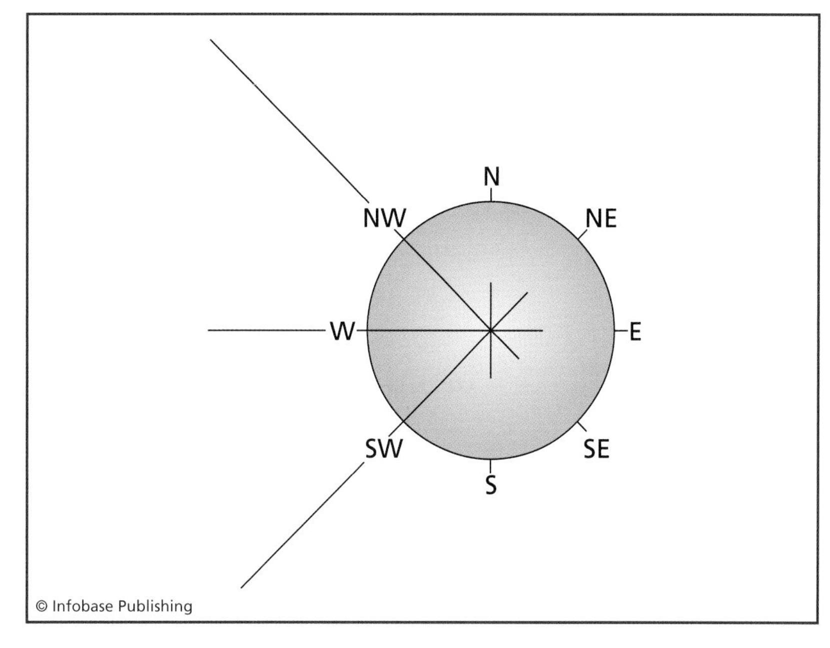

Each of the lines on a wind rose represents a direction from which the wind has blown during a reference period, often of one year. The length of each line indicates the frequency with which the wind has blown from that direction.

Herodotus (ca. 484–ca. 425 B.C.E.; see "Herodotus and his Travels" on pages 107–109) mentioned the use of the Sun and stars in navigation, and so do the Norse sagas that recount the exploits of Viking explorers.

The rising and setting of stars indicates east and west, and they also indicate north and south. Mariners observed that the stars appear to describe a circle around a certain point in the night sky. This was due to the Earth's rotation and so it was they who were turning and not the stars, but they did not need to understand that. They noted that wherever they were, the center of the star circle was due north of their position, and that a star lay almost at that center, as though it stood directly above the North Pole. They called this star Polaris. Today it is also known as the Pole Star, as well as the North Star. The angle between a line from the observer to the Pole Star and the horizon, known as the degree of latitude, indicates the observer's distance from the North Pole.

The Pole Star is not directly above the North Pole. Greek navigators noted that fact in about 320 B.C.E., although the two were slowly becoming more aligned. The North Star is more nearly above the North Pole today—about one degree away—than it was in their time. It will be closest in 2105. As the solar system moves around the Milky Way galaxy, the apparent positions of the stars change. In 3000 B.C.E. a star called Thurban stood above the North Pole, and about 5,000 years into the future the pole star will be Aldemarin. Sometimes no star occupies that position. In the Southern Hemisphere there is no single pole star at present, but a group called the Southern Cross indicates south.

WHICH WAY IS NORTH?

The points of the compass—north, east, south, and west—are not arbitrary directions. They refer to the way the Earth is oriented and turns on its axis. The rotational axis passes through the center of the Earth, emerging at the North and South Poles, but it is not quite that simple.

In about 280 B.C.E. the Greek astronomer Aristarchus (310–ca. 230 B.C.E.) noted something curious. The Earth takes one year to complete a single orbit of the Sun. Midwinter Day is the day on which the noonday Sun is lower in the sky than it is on any other day of the year, and the number of hours of daylight reaches a minimum;

Midsummer Day is the day when it is highest and the day is longest. The two midpoints between these dates are the spring and autumn *equinoxes*—so named because everywhere on Earth on those days the Sun is above the horizon for exactly 12 hours and below it for 12 hours, and thus day and night are of equal length. It follows, therefore, that a year can be counted as the time that elapses between two spring equinoxes or two autumn equinoxes—but when he recorded the positions of the stars at the same hour of the night on the equinoxes marking the start and end of a year, Aristarchus found they were different. The stars were not in the positions they had been in precisely one year earlier, and when he noted the times when they were in the same position they had occupied one year earlier he found he had measured a year with a slightly different length than the one given by the equinoxes. Aristarchus had discovered that the *sidereal year* (the year measured by the positions of the stars) differed from the *tropical year* (the year measured as the interval between equinoxes). The difference is small. Today the length of a tropical year is 365.242 days and that of a sidereal year is 365.256 days, so the sidereal year is 20 minutes and 9.6 seconds longer than the tropical year.

Although the difference is small, it is significant. In about 150 B.C.E. another Greek astronomer, Hipparchus (ca. 190–ca. 120 B.C.E.), found that this difference in year lengths was linked to gradual changes in the dates of the equinoxes. He observed that the equinoxes measured by the positions of the stars are moving westward along the *plane of the ecliptic* with respect to the equinoxes measured by the tropical year. That is, on successive spring or autumn equinoxes the stars are slightly to the east of the positions they occupied at the same time one year earlier. The plane of the ecliptic is an imaginary disk with the Sun at its center and its circumference marked by the Earth's orbital path. (Because this movement of the equinoxes was also recorded in a Sanskrit text from the 12th century C.E., we know that Indian astronomers discovered it as well.) At present the equinoxes are moving by one degree of arc every 71.6 years; in approximately 25,765 years they will have returned to their present positions. This phenomenon affects the world's climates, because it affects the dates on which Earth is at its closest *(perihelion)* and farthest *(aphelion)* from the Sun. At present, Earth is at perihelion on January 4 and at aphelion on July 4.

The dates of the equinoxes change because the spinning Earth behaves like a gyroscope. The Earth is not perfectly spherical. Its diameter at the equator is about 27 miles (43 km) greater than its pole-to-pole diameter. The gravity of the Sun, Moon, and, to a lesser extent, Jupiter exert a force on the Earth's equatorial bulge. Because the Earth is spinning and its axis is not at a right angle to its plane of orbit, that force causes the rotational axis to shift in a direction at 90° in the direction of rotation. The effect is called *precession* and when applied to the Earth it is the *precession of the equinoxes.*

The Earth's rotational axis is not at right angles to the plane of the ecliptic. Earth is tilted by an amount that varies over about 21,000 years from 22.1° to 24.5° and then back again over another 21,000 years. At present the axial tilt is 23.45°. It is the axial tilt that produces the Earth's seasons, because as the planet orbits the Sun, first one hemisphere is tilted toward the Sun and then the other. The larger the angle of tilt, the more extreme are the climatic differences between summer and winter in high latitudes.

A navigator can find the direction of north by observing the shadows cast by the noonday Sun, or by locating the Pole Star at night, but the concept of north and south is rather trickier than it might seem—and there is another north. The Earth has a magnetic field that is shaped like the field of a bar magnet. Physicists are not certain of all the details of how the field is generated, but they are confident that it results from the combined effects of the composition and rotation of the planet. The Earth's inner core is solid and made from iron and nickel. The thick outer core surrounding the inner core is also made from iron and nickel, but it is liquid. As the Earth turns on its axis, the liquid outer core drags slightly. The different rates of rotation of the inner and outer core cause a dynamo effect that generates an electric field. The electric field generates a magnetic field that is at right angles to the electric field. The magnetic field has a north pole and a south pole, and a line between them is at an angle of about 11.3° to Earth's axis of rotation. The magnetic south pole is close to the geographic North Pole, so the north pole of a bar magnet points toward it. However, to avoid confusion, what is physically a south pole is called the North Magnetic Pole. The magnetic poles wander. During the 20th century the North Magnetic Pole was moving northward by about 6 miles (10 km) a year, but in the early 21st century it accelerated to about

25 miles (40 km) a year. At present the North Magnetic Pole is at 82.7°N, 114.4°W and the South Magnetic Pole is at 64.5°S, 137.9°E.

William Gilbert (1544–1603), an English physician and physicist, was the first person to realize that the Earth is magnetic. He spent a large amount of his own money researching magnetism and in 1600 he wrote a book, *De Magnete, Magnetisque Corporibus, et de Magno Magnete Tellure* (Concerning magnetism, magnetic bodies, and the great magnet Earth), in which he described how to make a bar magnet from iron and asserted that the Earth behaves like a bar magnet. He also showed that contrary to popular belief, *lodestone*—also called magnetite, a naturally occurring iron-oxide mineral ($Fe^{2+}Fe^{3+}{}_2O_4[FeO.Fe_2O_3]$) with magnetic properties—does not cure headaches or improve health. If Earth behaves like a bar magnet, then a small bar magnet that is able to move freely will align itself with the Earth's magnetic field and will point north and south.

INVENTION OF THE MAGNETIC COMPASS

William Gilbert discovered that the Earth has a magnetic field, but he did not invent the magnetic compass. During the Warring States Period of Chinese history (475–221 B.C.E.), craft workers—possibly workers skilled in making jade ornaments—invented what they called *Si Nan,* or South Pointer. It comprised a piece of lodestone fashioned into the shape of a ladle to represent the Big Dipper constellation that points to the Pole Star. The bowl of the ladle was placed in a hollowed plate of polished bronze, where it could turn freely. When it was allowed to do so, the handle of the ladle pointed south. The device had little practical value, however, because lodestone is easily demagnetized.

There are three ways to make a small magnet from a piece of iron or steel: stroking it against lodestone; heating it to white heat, aligning it north and south, and then, while keeping it aligned, plunging it into cold water; and aligning the heated metal north and south and hammering it repeatedly. These methods work by allowing the iron molecules to align themselves with the Earth's magnetic field, then locking them in that alignment by cooling the metal so they lose the freedom to move. During the Northern Song dynasty (960–1127 C.E.) Chinese factories were using all three methods to manufacture magnets. Their products included a simple compass called the south-pointing fish, designed to help travelers find their way at night. It consisted of a small piece of thin iron sheet in the shape of a fish that had been magnetized by heating and was placed on the surface of water in a bowl. Surface tension allowed it to float, and it oriented itself to point north and south. It was a very weak magnet, however, so it was a toy rather than a useful piece of equipment. In other versions, a bar magnet was fastened to the underside of a floating wooden fish, and magnetized needles were hung from single-strand threads.

It is not certain when magnets were first used as serious aids to navigation. There are records of their use in the early 12th century in China, the late 12th century in Western Europe, around 1220 in the Arab world, and by 1300 in Scandinavia. At first, sailors probably used them to check the wind direction at times when clouds obscured the sky. By the 13th century, European compasses were mounted on a wind rose (see "Discovery of the Pole Star" on pages 80–82), allowing the navigator to read the ship's heading—approximately the same as the wind direction in a square-rigged ship. A convention arose (and continued until quite recently) whereby a fleur-de-lis indicated north and the line pointing east terminated in a cross because it was the direction of the Holy Land.

Ships are not stable platforms for a compass that floats on water, however, and a major development came in the 16th century. A marine compass consisted of a bar magnet fixed to the underside of a card painted with the points of the compass and floating in a bowl of water or oil. William Barlow (died 1625) designed a compass in which the bowl was attached by gimbals to a wooden frame. The gimbals allowed the bowl to remain horizontal regardless of the ship's motion, making it much steadier and more accurate. Barlow was a clergyman who was educated at Balliol College, University of Oxford, and eventually became chaplain to the Prince of Wales and, in 1615, archdeacon of Salisbury Cathedral. He combined his clerical duties with his researches into magnetism and he corresponded with William Gilbert. The illustration on page 87 shows the compass Barlow designed in 1597, and points out its other important feature. By the 16th century navigators knew that the North Magnetic Pole and the geographic North Pole were not in the same place, and by using astronomical observations of true north they were able to measure the angle between them, called the *magnetic variation.* Obviously,

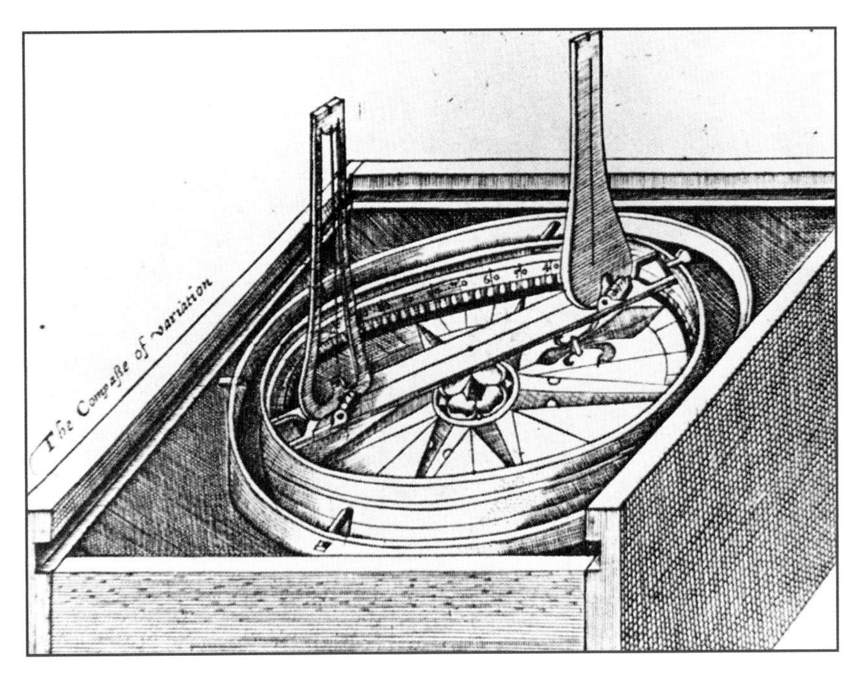

A magnetic compass designed in 1597 by Archdeacon William Barlow. The compass card, showing the points of the compass, is mounted on gimbals so that it remains horizontal regardless of the motion of the ship. It is called a variation compass, because—provided the user knows the angle between geographic and magnetic north—the compass can be adjusted to show geographic north. *(Science Photo Library)*

that angle varied with longitude, but it was not difficult to compile a table showing the magnetic variation at each angle of longitude. Provided he knew the longitude (see "Measuring Longitude" on pages 99–102), Barlow's "variation compass" allowed the navigator to preset the magnetic variation so that a line above the compass card showed geographic north to one side of magnetic north.

The magnetic compass has remained in use to the present day and as recently as the 1940s most aircraft were fitted with compasses consisting of cards floating on a liquid. They were difficult to use, however, because the compass lags behind the motion of a ship or airplane, then takes time to settle sufficiently to give a reliable reading. When steel replaced wood as the material for building ships, magnetic compasses had to be shielded from the distortions due to their proximity to the metal, and they also had to be shielded from the magnetic field generated by aircraft electrical equipment. In 1908, the German inventor Hermann Anschütz-Kaempfe (1872–1931) patented the gyrocompass in Germany and the American inventor Elmer Ambrose Sperry (1860–1930) patented it in the United States. A gyrocompass is a gyroscope set to indicate true north and is not based on a magnet.

PILOT BOOKS

Merchants may have taken commercial risks, but few of them were adventurers, and the ships they owned or chartered sailed mainly along traditional routes, often between ports along a single stretch of coastline. Navigators needed to recognize prominent land features such as inlets, headlands, and high cliffs in order to keep track of their progress and when they arrived at their destination they needed to know how to enter the harbor safely. Large ships often took a pilot on board to guide the vessel into port, but smaller ships could not afford the expense. Their navigators had to be able to find their own way, and they passed on their knowledge from one generation to the next. There were no charts or maps to guide them, and for centuries there were no written instructions of any kind.

That situation began to change when documents, and eventually books, of sailing instructions began to appear, probably about 2,400 years ago. A pilot book or periplus described the coastline between ports as a navigator would see it and it gave instructions for sailing from one port to the next. It explained the direction to sail in terms of the prevailing wind direction and the distance as the number of days sailing, and for each port it described the entrance, sea currents encountered on the approach, and safe anchorages.

The earliest pilot book described a voyage made by Hanno the Navigator, a Carthaginian (Phoenician) explorer who lived around 500 B.C.E. and who claimed to have led a fleet round the coast of Africa from Carthage (near modern Tunis) through the Pillars of Hercules (Straits of Gibraltar) and perhaps as far as modern Cameroon—he described what may have been a volcano resembling Mount Cameroon, which is 13,370 feet (4,075 m) high and clearly visible for some distance offshore. The mountain may not have been a volcano at all, but a mountain on which a bush fire—perhaps a signal fire—was burning. Hanno gave it the local name, calling it the Chariot of the Gods. If it was a hill with a fire rather than a high volcano, the Chariot of the Gods could have been in Sierra Leone. Scholars agree that Hanno reached that far, but doubt that he could have traveled as far as Cameroon.

When Hanno returned to Carthage he is said to have hung up a tablet describing his voyage in the temple of Ba'al Hammon, the principal god of the Phoenicians. The tablet has not survived, but before it was lost it was translated into Greek and given the title *Periplus*. Modern scholars believe the account is genuine and that the voyage took place in about 425 B.C.E. with the aim of establishing new Carthaginian colonies. According to the *Periplus,* Hanno "sailed with 60 ships of 50 oars each, and a multitude of men and women to the number of 30,000, and provisions and other equipment."

Hanno described the countries they visited and their people and animals, including "river horses" or hippopotamuses—Greek *hippo,* "horse," and *potamon,* "river." They also recruited interpreters to help them converse with the people they met. The expedition did not confine itself to following the coast. When it arrived at the mouth of a large river the ships sailed upstream to explore the interior. At the farthest point of their journey, the *Periplus* describes in the following words what the expedition found.

> On the third day of our departure thence (from the Chariot of the Gods), having sailed by streams of fire, we came to a bay which is called the Southern Horn. At the end of this bay lay an island like to that which has been before described. This island had a lake, and in this lake another island, full of savage people, of whom the greater part were women. Their bodies were covered with hair, and our interpreters called them Gorillas. We pursued them, but the men we were not able to catch; for being able to climb the precipices and defending themselves with stones, these all escaped. But we caught three women. But when these, biting and tearing those that led them, would not follow us, we slew them, and flaying off their skins, carried these to Carthage. Further we did not sail, for our food failed us.

In 1836, the American missionary and naturalist Thomas Staughton Savage (1804–80) was sent to Liberia. While there Savage acquired the skull and other bones of an ape that was then unknown in America and Europe. He described it at a meeting of the Boston

Society of Natural History, giving it the name of Hanno's "savage, hairy women." He called it *Troglodytes gorilla;* it is now known as the western gorilla *(Gorilla gorilla).* This was a reasonable identification, because it is likely that Hanno's party had encountered apes, probably gorillas, rather than a tribe of humans.

MEASURING SPEED

Reporting distances as the time it takes to cover them makes obvious sense. In many cultures, someone who asks the distance to the next village will be told the answer as the number of days it will take to walk there. This sounds practical, but there is a problem: Not everyone walks at the same speed. The problem is greater for sailing ships, because it is the strength and direction of the wind that determines their speed, and the wind is highly variable. Consequently, navigators needed some method for measuring the speed of the ship.

The device that measures the speed of a ship is called a *log,* because the very first version consisted of just that: a wooden log. The navigator would stand at the bow of the ship, throw the log over the side, then walk to the stern and note how long it took for the ship to sail past the log, which remained stationary in the water. The navigator knew the precise length of the ship, of course, so it was simple to divide the time that elapsed by the distance traveled to arrive at the speed: $S = D/T$, where S is speed, D is distance, and T is time. The log was sometimes called the *Dutchman's log,* although the first use of that term appears to date from 1623 and the method was certainly used in ancient times.

In 1637, the English mathematician and surveyor Richard Norwood (ca. 1590–1675) proposed an improvement in his book *The Seaman's Practice, containing a Fundamentall Probleme in Navigation experimentally verified, namely, touching the Compasse of the Earth and Sea and the quantity of a Degree in our English Measures.* On June 11, 1633, Norwood measured the height of the noonday Sun at London, and on June 6, 1635, he measured it at York. This gave him the latitudes of the two cities. He then measured the distance between London and York as 367,176 feet (111,915 m), and from that he calculated the length of one degree of latitude. The nautical mile is approximately equal to the length of one minute of arc of latitude, or one-sixtieth of the length of a degree of latitude. An international standard length for the nautical mile was established in 1929, as 1,852 meters (6,076.4 feet).

Norwood's improvement to the log was to tie knots in a line at intervals of 50 feet (15 m), fasten one end of the line to the log, and to use a 30-second sandglass to measure the time. The navigator counted the number of knots that were paid out before the sandglass was emptied, and that number was the ship's speed in *knots*—a measure of speed of one nautical mile per hour.

In time, the *chip log* replaced the log of wood. This was a more accurate device, consisting of a quarter-circle piece of wood held by a bridle of three lines from the corners that was attached to the knotted line. The illustration below shows the arrangement. When the chip log was thrown from the ship's stern it acted as a drogue (sea anchor), remaining stationary in the water while the ship moved away from it.

The next development occurred in 1688 when Humphry Cole, an English instrument maker, invented the *taffrail log,* also called the

The quarter-circle wooden board is attached to a line divided by knots into measured lengths and wound onto a reel. The board would be thrown from the ship's stern and the line allowed to run freely from the reel for a measured interval of time.

patent log. This was a metal device shaped like a torpedo with vanes at the rear end that were free to rotate as the log was towed through the water. It was lowered from the stern and the ship's speed was calculated from the number of turns made by the vanes in a given time.

All ship captains are required to keep a record of the voyages they undertake. Since the early 19th century the book containing such a record has been known as a *logbook,* sometimes shortened to log. The use arose because captains would report the distance the ship had covered in a day as having been measured "by the log." Logbooks are also maintained for all aircraft movements.

COUNTING TIME

Speed is the rate at which a moving body changes its position. To determine it accurately, therefore, the navigator had to measure two quantities: distance traveled and time elapsed. The log, with its later improvements, proved a reliable device for measuring distance traveled, but measuring time was more difficult.

It was probably the Sumerians, about 5,500 years ago, who first divided the day into intervals and invented the ancestor of the sundial to measure the passage of time. A Sumerian sundial was an *obelisk*—a tall stone column that tapers upward and ends in a point. The direction of its shadow followed the movement of the Sun, indicating the time of day. About 3,500 years ago the Egyptians invented a sundial that was small enough to be carried about. It was shaped like a T-square, but with the crossbar of the T set upright so it cast a shadow on the long arm of the T, as shown in the illustration on page 93. At sunrise the device was set on a level surface with the crossbar facing to the east. The shadow cast by the cross-bar then marked the morning hours. At noon the device was turned around, so the crossbar faced to the west and its shadow marked the afternoon hours. The Egyptians divided the day into 10 hours, plus two twilight hours at dawn and dusk. Obviously, sundials do not work at night, but about 2,600 years ago the Egyptians invented an instrument that could measure the movement of the stars in relation to the Pole Star.

Sundials are of little use when clouds hide the Sun, but there is another way to approach the problem. Any process that moves at a regular, steady pace provides the basis for making a clock, and falling

At sunrise the sundial is set with the upright bar facing east; the shadow then marks the morning hours. At noon it is turned around so the bar faces west; its shadow then marks the afternoon hours.

water was the first to be tried. The Egyptian pharaoh Amenhotep I (1525–04 B.C.E.) owned a water clock—one was found in his tomb. It measured the passage of time by both day and night regardless of the weather, provided someone remembered to keep its reservoir topped up. The Greeks called the water clock a *clepsydra,* which means "water thief," and equipped it with a system of gears and levers that made it move a pointer on a dial. Roman inventors linked water clocks to devices that struck gongs or bells, making the first alarm clocks. Water clocks of increasing sophistication remained in use until the Middle Ages, and to this day a few are manufactured, but more as novelties or public art than serious timepieces. No one knows when water clocks were invented. Some historians believe they may have been in use in China 6,000 years ago.

It is possible to make a candle that burns at a very steady rate. If it is marked with a scale such a candle can serve to measure the passage of time. Incense can do the same. Candles and incense have often been used to mark time in temples and churches.

None of these devices can be used at sea, because they work reliably only if they are on a steady base and ships are in constant motion. Consequently, seafarers used sandglasses. These were probably invented in the third century C.E. and are first recorded in Alexandria, Egypt. They are robust, do not freeze, and are fairly accurate. The length of time they measure depends on the quantity of sand they hold. The most widely used sandglasses hold enough sand to measure the passing of one hour and thus are called hourglasses, but sandglasses can measure any other time interval equally well; egg-timers, for example, measure three minutes. Sandglasses measuring different intervals were sometimes set side by side in a wooden frame.

Sandglasses were carried on ships from the 11th century and a member of the crew had the job of turning them. Ferdinand Magellan (1480–1521; see "Ferdinand Magellan, from Atlantic to Pacific" on pages 132–135) had 18 sandglasses on each of his ships. At sea, working shifts, called watches, divide the 24 hours into intervals, ensuring that duties are shared fairly among the crew. The ship's bell marks the passing of time and the changing of watches. Sandglasses are perfectly adequate for timing the ringing of the bell. They are not adequate, however, for measuring *longitude*—the angular distance from a line passing through Greenwich, England, and through the geographic North and South Poles (see "Measuring Longitude" on pages 99–102). That requires much greater accuracy than any sandglass can achieve and it became possible only with the invention in the 18th century of a robust but extremely precise clock (see "John Harrison and his Time-Keeper" on pages 102–105).

The word *clock* is derived from the medieval Latin word *clocca* ("bell"); strictly speaking, a timepiece should be called a clock only if it chimes or otherwise indicates the time audibly. The name Big Ben actually refers to the bell that sounds the hours in the clock on the tower above the Houses of Parliament in London, although it is commonly applied to the clock itself. The first mechanical clocks appeared in the 14th century. They were large and installed on towers in some Italian towns.

Galileo Galilei (1564–1642) drew a design for a clock based on the regular movement of a pendulum, but the Dutch physicist Christiaan Huygens (1629–95) was the first person to make one. It was accurate to within one minute a day and later improvements reduced this to less than 10 seconds. In about 1658 the English physicist Robert Hooke (1635–1703) invented the spring and balance-wheel mechanism that could replace the pendulum, but Huygens made the first clock to use this mechanism in about 1675. Spring and balance-wheel watches were accurate to within about 10 minutes a day, so they were unsuitable for navigation, but they were probably as reliable as sandglasses. It was simple to check them each day at noon, and their convenience made them popular. Clocks and watches began to displace sandglasses.

CROSS-STAFF AND SEXTANT

Measuring the speed of the vessel was important, but that information was of navigational use only if the navigator was also able to measure the position of the ship at regular intervals. The first step was to measure the ship's latitude. At night the angular distance between the observer and the Pole Star—how high the star is in the sky—is equal to the latitude of the observer. It is easy to see why this must be so. The Pole Star is such a long distance from Earth that for practical purposes all straight lines from the surface of the Earth to the Pole Star are parallel. The Pole Star is directly above the North Pole. Consequently, an observer at the equator will see the Pole Star just touching the horizon, at an angle of 0° to the horizon; the equator is at latitude 0°. An observer at the North Pole will see the Pole Star directly overhead, at an angle of 90° to the horizon; the North Pole is at latitude 90° N. At any point between the equator and the North Pole the angular height of the Pole Star above the horizon must be the same as the angle of latitude.

The earliest European device for measuring the angle between any celestial object and the horizon was the *cross-staff*, also known as the Jacob's staff, probably invented by the Jewish astronomer Jacob ben Machir ibn Tibbon (ca. 1236–ca. 1304), who lived in Provence, France. It consisted of a staff, marked with graduations of length, and a cross-piece, also called a transom or transversal, that was able to move along the staff. To use the device, the end of the staff was placed against the cheek just below the eye, with the cross-piece vertical. The user moved the staff until one end of the cross-piece was just touching the horizon, then slid the cross-piece along the staff until its other end touched the celestial object. The position of the cross-piece on the staff could then be converted into the angle of elevation by trigonometry: If the distance along the staff between the user's eye and the cross-piece was AC, the half-length of the cross-piece was BC, and the angle of elevation was A, then $\tan A = BC/AC$. In fact, there were tables showing the angle for different values of AC and BC. Similar instruments appeared in China during the 11th century and in India during the 12th.

At first the cross-staff was an astronomical instrument and not used for navigation. A German mathematician, Johannes Werner (1468–1522), pointed out in 1514 that it could be used to calculate latitude, but it was not until the 1550s that the English astronomer and mathematician John Dee (1527–1608) introduced it to English navigators.

The cross-staff was difficult to operate on a ship. The navigator had to hold the device steady and continually shift his gaze in order to align the ends of the cross-piece correctly, so he was trying to look in two directions at once while the ship was moving beneath his feet. Improvements were made over the years, one of which was to graduate the staff directly in degrees, which removed the need for a trigonometric calculation or reference to printed tables. But there was a more serious difficulty. The Pole Star is not especially bright; a thin veil of cloud will obscure it. It was much easier to calculate latitude from the elevation of the noonday Sun. To do this, however, the navigator had to look directly at the Sun, which is difficult, painful, and potentially extremely dangerous—it can cause blindness. So, to avoid this risk, first a piece of smoked glass was added to reduce the glare, and then the cross-staff was turned around and became the *backstaff.*

To use the backstaff, or back-quadrant, the navigator faced away from the Sun. The vertical cross-piece had become an arm that cast a shadow onto a horizon vane at the far end of the staff. The navigator looked down the staff and moved the arm back and forth until the Sun's shadow was aligned with the horizon, seen through a horizontal slit in the horizon vane.

There were many versions of the backstaff and gradually it developed into the *quadrant,* which came into use during the 17th century.

Again, there were several versions, including ones with a horizon vane for use in daytime. A quadrant was a quarter-circle graduated in degrees, with a moveable arm fixed to the center of the circle. While holding the upper edge of the quadrant horizontal, the navigator looked along the arm, aligning it to a celestial object.

During the 18th century the *octant*—one-eighth-circle—replaced the quadrant as a navigation instrument. In about 1699, Sir Isaac Newton (1643–1727) designed a quadrant with a 45° arc that used two mirrors and had a sighting telescope mounted on the side, but the design was not published until 1742. Meanwhile, in about 1730, both the English mathematician John Hadley (1682–1744) and the American optician Thomas Godfrey (1704–49), working independently, invented the octant. This had a 45° graduated arc, a sighting telescope, and three mirrors. The index mirror, with a shade that could be moved forward to prevent excessive glare, was situated beside the telescope eyepiece and the two smaller horizon mirrors at the opposite end, in the line of sight of the telescope. The horizon mirrors showed a reflection of the image in the index mirror. The user directed the sighting telescope at a distant object and could then see that object and beside it the horizon mirror showing a reflection of the image in the index mirror. Moving an index arm tilted the index mirror until it reflected a second object, which the user saw in the horizon mirror beside the first object. The position of the index arm on the arc then indicated the angle between the objects. Setting the two horizon mirrors at right angles to each other and remounting the sighting telescope made it possible to compare objects that were up to 180° apart.

The next step was to enlarge the octant until its arc spanned one-sixth of a circle (60°). It was then a *sextant.* Sextants came into use in the latter part of the 18th century and navigators continued to rely on them until the Global Positioning System (GPS) rendered celestial navigation redundant. The illustration on page 98 shows a modern sextant, with the parts labelled.

The navigator sees through the telescope the object at which the instrument is directed and beside it the image in the horizon mirror, which is a reflection of the image in the index mirror. In some sextants the horizon mirror covers half of the field of view, while in others the mirror is half-silvered. In either case the mirror image appears beside the horizon or celestial object at which the sextant is pointed.

Moving the index bar moves the index mirror and the navigator adjusts this until the two images are side by side. The micrometer allows for fine adjustments and the vernier scale on the micrometer

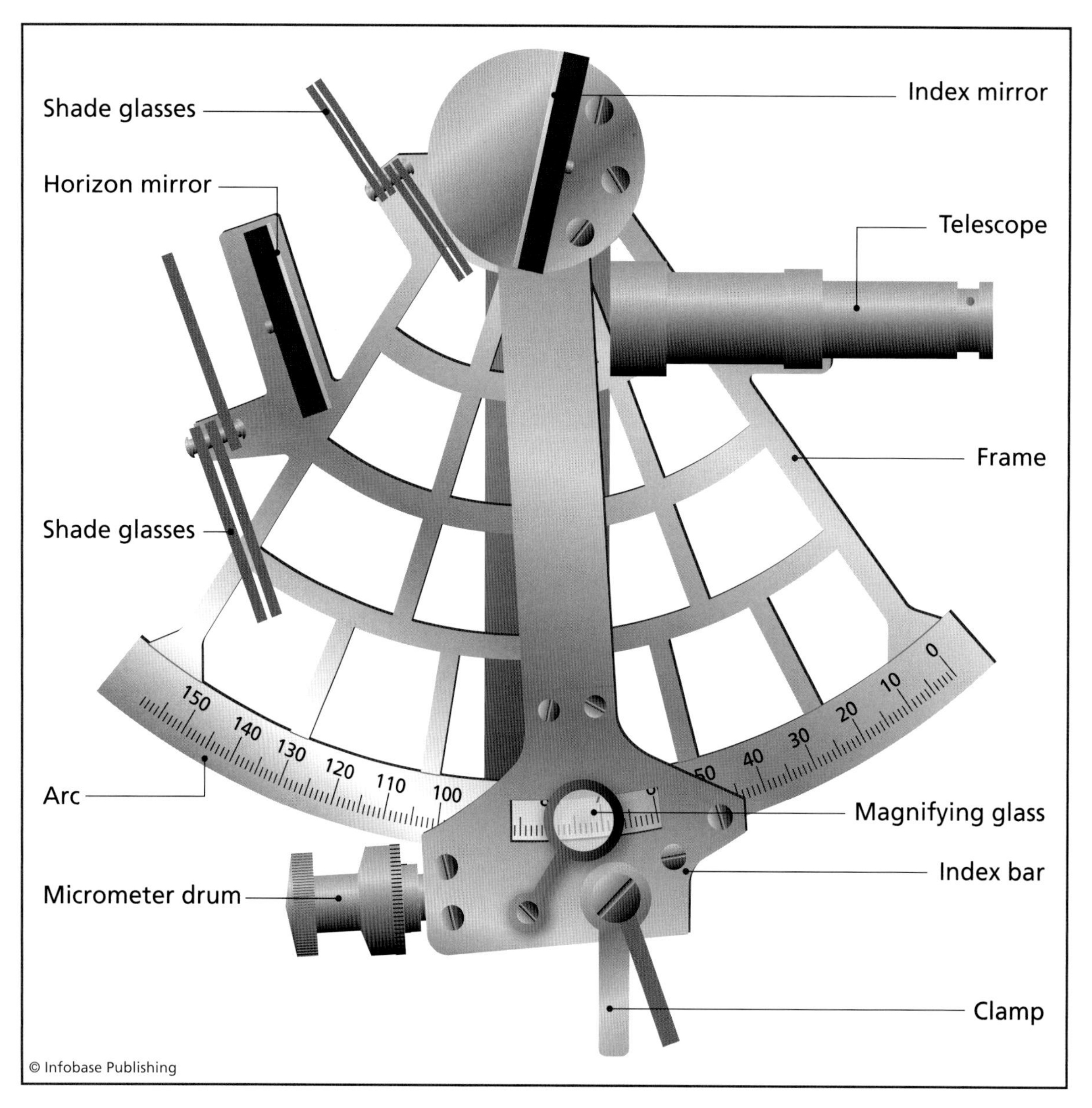

A modern sextant, used to measure the height of a celestial object above the horizon. A vernier scale on its micrometer allows it to measure angles to an accuracy of 0.2 minute, equivalent to approximately 0.2 nautical mile (0.23 mile [0.37 km]).

drum allows measures to within about 0.2 of a minute of arc (1/300th). There are shades that can be moved over the mirrors to reduce glare and a magnifying glass allows the scale to be read with greater precision.

There are times when the horizon is not clearly visible due to haze or darkness, although the sky overhead is clear. At such times the navigator can attach an artificial horizon (see sidebar on page 100). The illustration at right shows the type of artificial horizon that is used in aircraft. The type used on a sextant comprises a small spirit level—a tube partly filled with a liquid so there is a bubble—that is visible through the horizon mirror.

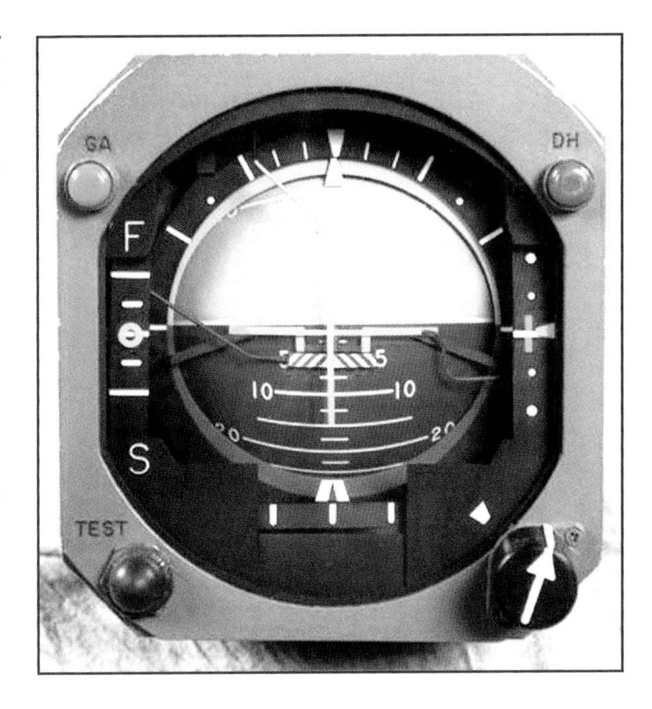

On an aircraft instrument panel the artificial horizon shows the attitude of the aircraft, i.e., whether its nose is above or below the horizon and whether its wings are level or it is banking. *(NASA/Ames Research Center)*

MEASURING LONGITUDE

As a ship prepared to leave port, at one time the navigator would carefully measure the latitude and write it down. When it was time to set a course for home, the ship would head north or south until it reached the latitude of the port from which it had departed. Then it would head east or west, maintaining the same latitude, until it reached home. It was a simple method that worked, even though it often increased the length of a voyage. If the navigator could have known the distance of the ship to east or west of its destination he could have set a more direct course. But to do that, he would have had to know the longitude. Measuring longitude is much more difficult than measuring latitude.

Amerigo Vespucci (1454–1512), the Italian explorer (see "Amerigo Vespucci in South America and the Caribbean" on pages 127–129), found one way to determine longitude. In 1499 he learned by consulting an almanac that on August 23, at midnight or half an hour before that the Moon would be very close to the planet Mars. When he studied the sky he found that at midnight there was a distance of 3.5 degrees between them. The almanac referred to the sky over Europe, so he was able to calculate his ship's distance west of Europe. His result was very approximate and his method was of limited use,

ARTIFICIAL HORIZON

An artificial horizon is an instrument that shows the position or attitude of an observer in relation to the horizon. On a ship it usually consists of a spirit level—a glass tube partly filled with liquid to leave a bubble—that is viewed through the horizon mirror of a sextant. When the air bubble is at the center of the tube, the sextant is horizontal (i.e., parallel to the horizon).

In an aircraft, the artificial horizon is one of the basic flight instruments the pilot must use when flying in clouds. It is basically a gyroscope powered electrically, or by a fine jet of air directed by a pump at cups around the heavy rim of a wheel. The spinning gyroscope provides a stable platform that maintains its orientation regardless of the motion of the aircraft. The face of the instrument shows a horizontal line—the horizon—that remains motionless as a second line representing the aircraft moves in relation to it. This arrangement allows the artificial horizon to show the aircraft's attitude. If the nose is inclined upward, the line representing the aircraft moves above the horizon and it moves below the horizon if the aircraft nose is inclined downward. The distance between the two lines indicates the degree to which the nose is high or low. The wings are level if the two lines are parallel. If the aircraft banks, the line representing the aircraft will tilt to left or right by an angle that indicates the degree of banking.

because it depended on a rare astronomical event—the relative positions of the Moon and Mars.

Lines of longitude, called *meridians,* are imaginary lines around the Earth from pole to pole. Each meridian crosses the equator at right angles. They are measured in degrees from a reference meridian. For centuries each country used a reference meridian that crossed its own territory, usually passing through its capital city. That made reporting a ship's location by its latitude and longitude extremely confusing, and in October 1884, delegates from 25 nations met in Washington, D.C., for the International Meridian Conference, one of the aims of which was to resolve that confusion. The delegates agreed that henceforth the international reference meridian should pass through the Royal Observatory at Greenwich, near London, England. Since then the Greenwich meridian has been 0° longitude and all other meridians have been counted to the east and west of it.

Calculating longitude is possible because of Earth's rotation. Earth turns on its axis, completing one revolution in 24 hours, so in 24 hours Earth turns through 360°. In one hour, therefore, it turns through 15° (360 ÷ 24). Consequently, if it is noon in a particular place when it is 3 P.M. at Greenwich, then that place is three hours behind Greenwich and must be at longitude 45° (3 × 15). Because it is behind Greenwich it must be farther to the west, so it is at 45° W.

The principle is simple enough, but in order to apply it the navigator needs an extremely accurate clock that shows the time at Greenwich. Local noon is the time when the Sun reaches its highest point in the sky. The most accurate way to determine that is by the equal altitude method, using a quadrant or sextant. The navigator sets an arbitrary angle of elevation and turns the instrument to the east, notes the time when the Sun crosses that angle, turns the instrument to the west, and notes the time when the Sun crosses the angle a second time. Noon is precisely halfway between those times. Each time the Sun crosses the angle, the navigator must also note the Greenwich time. It is a simple matter to calculate longitude from a comparison between the local noon and Greenwich noon.

The problem was that no clock existed that was capable of keeping time with sufficient accuracy onboard a ship, and even a small error translated into a significant distance on the Earth's surface. It was intractable. Pendulum clocks were sufficiently accurate, but only provided they stood on a surface that remained steady. Clocks with a spring and balance wheel were manufactured, but they were simply not up to the task (see "Counting Time" on pages 92–95). Most scientists dismissed as impossible the idea that anyone would ever be able to measure longitude satisfactorily.

Finally, matters came to a head. Lacking a method for calculating longitude, navigators had to rely on *dead reckoning*—estimating their ship's position from the measured speed and direction traveled since the last known position. In October, 1707, a squadron of English ships returning from a battle sailed through fog for 12 days, estimating their position by dead reckoning. Their navigators were mistaken—though through an error in latitude rather than longitude—and four of the ships foundered on rocks off the Isles of Scilly, with a loss of 2,000 men. In 1713 two mathematicians suggested a way to determine longitude. They proposed anchoring ships at intervals along the shipping lanes and having them regularly fire shells designed to explode at a height of more than one mile (1.6 km). The navigator on any ship within the area could measure the time that elapsed between seeing and hearing the explosion and work out from that the position of the ship in relation to the known longitude of the ship that fired the shell. The scheme would never have worked, but it won wide popular support. Then a number of ship captains petitioned Parliament, proposing that the government offer a prize for a practical solution. That was enough, and after much debate Parliament acted by passing the Longitude Act.

The *Journal of the House of Commons* reported the following for its entry on June 11, 1714:

> *Resolved, Nemine contradicicente**, that the House doth agree with the Committee in the said Resolution, That a Reward be settled by Parliament upon such Person or Persons as shall discover a more certain and practicable Method of ascertaining the Longitude, than any yet in Practice; and the said Reward be proportioned to the Degree of Exactness to which the said Method shall Reach.
>
> *No one disagreeing.

The Act received the Royal Assent from Queen Anne on July 20, 1714, less than two weeks before she died. John Harrison (1693–1776) rose to the challenge.

JOHN HARRISON AND HIS TIME-KEEPER

In 1714, under the terms of the Longitude Act, the British government offered prizes of £10,000 for a timepiece accurate enough to determine longitude to within 60 nautical miles (69 miles, 111 km); £15,000 if it could determine longitude to within 40 nautical miles (46 miles, 74 km), and £20,000 if it could determine longitude to within 30 nautical miles (34.5 miles, 55.5 km). In today's money these prizes would be worth approximately £1.3 million ($2 m), £2 m ($3.3 m), and £2.6 m ($4.2 m), respectively. The Longitude Prize was to be administered by the Commission for the Discovery of the Longitude at Sea, more popularly known as the Board of Longitude. The English clockmaker John Harrison decided to compete.

John Harrison was born on March 24, 1693, in the village of Foulby, near Wakefield, Yorkshire; in about 1700 the family moved to the village of Barrow upon Humber, in North Lincolnshire. His father was a carpenter and mechanic, and John followed in his footsteps, becoming increasingly skilled at making measuring instruments and mechanical devices. He had a particular interest in making clocks and built his first longcase clock in 1713. It was made entirely from wood and was the first of several wooden clocks that Harrison made, three of which have survived.

In 1718 Harrison married Elizabeth Barrel. They had a son, John, who was born in 1719. Elizabeth died in 1726, and Harrison married Elizabeth Scott. They had a son, William, and daughter, Elizabeth. John, Harrison's first son, died in 1737.

Pendulum clocks depend critically on the length of the pendulum and until that time their accuracy was limited by the fact that metal pendulums expand and contract with changes of temperature. In 1726, Harrison exploited the different rates of expansion of two different metals to make a compound pendulum that remained the same length regardless of the temperature. By 1728 he had made his first marine chronometer.

In 1730 Harrison went to London to seek financial backing for his attempt at the Longitude Prize. He showed designs for a suitable chronometer to the astronomer Edmond Halley (1656–1742), who referred him to George Graham (1673–1751), one of the country's most distinguished horologists. Graham loaned Harrison the money he needed from his own pocket.

By 1735 Harrison had constructed his first chronometer and he demonstrated it to the Royal Society. They recommended it to the Board of Longitude and in 1736 the chronometer underwent its first sea trial on a voyage, out to Lisbon on HMS *Centurion* and back on HMS *Orford.* The chronometer performed well, but the board required a transatlantic voyage to test it fully and awarded Harrison £500 to continue his work. He completed a second chronometer in 1737, but found it unsatisfactory. He asked for and received a further £500 and used it to build a third chronometer, working on this one from 1740 until 1759, only for it to prove insufficiently accurate to meet the board's specification.

In 1753 Harrison commissioned a watch from a London watchmaker. He intended this for his personal use, but realized that certain improvements could convert it into an accurate chronometer. In 1755 he asked the Board of Longitude for more money, partly to continue work on his third design and partly to make two watches, one small enough to be carried in the pocket and the other larger. His fourth marine chronometer was circular in shape, five inches (13 cm) in diameter, and weighed just over three pounds (1.45 kg). That was the instrument sent for the full sea trial. Harrison's son William carried the chronometer on HMS *Deptford,* departing for the West Indies on November 18, 1761, and arriving in Jamaica on January 19, 1762. The chronometer was found to be only 5.1 seconds slow, equivalent to a little more than one minute of longitude, but the board was not satisfied.

On March 28, 1764, William set off on board HMS *Tartar* for a second trial, this time bound for Barbados and accompanied by the astronomer Nevil Maskelyne (1732–1811). The voyage took 47 days, and at the end the chronometer was wrong by 39.2 seconds, equal to less than one minute of longitude. That was good enough to qualify for the £20,000 prize, but the board held that the result was pure chance, and they demanded that other clocks be made to the same design. They agreed to pay Harrison £10,000 for the details of his design and a further £10,000 when more chronometers had been made with the required accuracy. At first Harrison refused, but after several weeks he and William gave in, and in August 1765 a team of six experts declared themselves satisfied that they were in possession of the complete design.

John Harrison's fifth chronometer, which he completed in 1770, was little larger than a pocket watch. It gained 4.5 seconds in 10 weeks. King George III tested the timepiece personally. *(Science Photo Library)*

The Board of Longitude then demanded that four chronometers be made to the same design and handed over to them. Harrison agreed and recommended Larcum Kendall (1721–95) as a watchmaker competent to make them. Finally, he received his first £10,000.

To qualify for the second installment, the board insisted that Harrison should make at least two more of the chronometers himself, meanwhile sending the original to be tested at the Royal Observatory. Harrison, by then in his 70s, worked with William to make the instruments. On January 31, 1772, news of the way the board was treating Harrison reached King George III and Harrison was summoned to meet him. The king was shocked at what he learned and declared "By God, Harrison, I will see you righted!" The king tested Harrison's fifth chronometer personally in 1772 and found that it performed superbly, but the board refused to recognize this test as valid. The illustration on page 104 shows the fifth chronometer, tested by the king. The Harrisons then petitioned Parliament, and in June 1773 an Act of Parliament recognized that they had solved the longitude problem and Harrison was awarded £8,750.

On his second (1772–75) and third (1776–79) voyages James Cook (1728–79) carried the first of Kendall's copies of Harrison's chronometer and recorded its accuracy in his logbook. John Harrison died on March 24, 1776, at his home in London.

The Wanderers

Throughout history, each generation has produced individuals who were not content to stay at home all their lives. They were discontented, bored with village life, or simply curious. So they left the village to explore the world beyond its confines. Others did not leave home by choice but were compelled to do so. Whenever rulers made war they called on the wealthy landowners in their realms to supply them with soldiers. The landowners passed on the demand to their stewards and every community was required to contribute a stated number of men of fighting age. These conscripts marched away to distant battles from which many never returned. They saw other lands, met people who spoke differently, had different customs. But few of them recorded their experiences.

This chapter tells of a different kind of explorer, the kind who did record what he (it was usually a he) saw and heard. Their bravery was remarkable because they knew the risks they were taking when they set forth into completely unknown territory with no maps to guide them. They expected to encounter sea storms in their small, fragile ships, and they did not expect the people they met to be friendly.

The chapter begins with tales of ancient Greek and Roman travelers. It tells of medieval journeys from Europe to the Far East and of the great voyages of exploration that eventually encircled the Earth. Step by step, those journeys contributed information that accumulated to form a changing image of the world that was reflected in maps. The chapter considers a few of the most important historical developments in mapmaking.

HERODOTUS AND HIS TRAVELS

The Phoenicians traveled widely and saw many things, but there is one story they told that Herodotus (ca. 484–ca. 425 B.C.E.), one of the leading historians and travelers of the ancient world, dismissed as nonsense. The Phoenicians claimed that while sailing westward in African waters the Sun was to their right. That could not be, of course. Like every educated European, Herodotus knew that the Sun lay to the south, so it had to be on the left when sailing westward. The Phoenicians could not have made up this story. It meant they had crossed the equator and were sailing in the Southern Hemisphere, but although the Greeks knew that the Earth was spherical, at that time no one knew that the Sun stood overhead at the equator and would appear to the north rather than to the south to a person south of the equator.

Herodotus was most famous as a historian, in the modern sense of the word, and his major work, to which he devoted much of his adult life, was called *The Histories.* He wrote it in about 440 B.C.E., in the Ionian dialect of the Greek language. It comprised nine books, each named after one of the Muses—the nine daughters of Zeus and Mnemosynē (Memory), each of whom presided over a particular science or art. So far as he could, Herodotus collected information firsthand and checked its accuracy. That explains why he rejected a Phoenician claim he considered implausible.

The subject of his histories concerned the wars fought between the Achaemenid Empire of Persia and the city-states of Greece in the fifth century B.C.E. and the events leading up to them. In order to construct his story of those events, Herodotus visited Upper Egypt, Babylon, Greece, Scythia (now southern Russia), all the countries making up Asia Minor (now Turkey), Italy, part of the coast of North Africa, and a number of islands, possibly including Crete. He traveled inside the Persian Empire using the official staging posts that formed part of the postal service, each post one day's journey from the next. Herodotus had completed his travels by the time he was 40.

Although he was a keen observer and reliable recorder, Herodotus spoke only Greek. Consequently, he had to rely on local interpreters as he journeyed through the Persian Empire, where many languages were spoken. Translation errors may have made some of his descriptions unreliable, but in the main, Herodotus has proven an important source of knowledge about the Mediterranean and Black Sea regions at that time. In his Book II (named after the muse Euterpe), he described the way Egyptians sailed the Nile (see "Egyptians on the Nile" on pages 2–4); he knew that the annual Nile flood followed the melting of the snows farther south, although he was deeply puzzled about how there could possibly be snow in what he knew was the hottest part of the world. He noted that Egyptian women conducted business in the markets and that men worked as weavers. He mentioned the Egyptians' calendar and their understanding of surveying. Herodotus also described the animals of Egypt, including the crocodile, hippopotamus, and otters, as well as fabulous beasts such as the phoenix and winged snakes. In Book III (Thalia) he described the cultures of India and Arabia, although he never visited these countries.

In Africa, Herodotus described the indigenous people he knew as Libyans and Ethiopians, and the immigrant peoples, who were Greek and Phoenician. He found people who dressed as Libyans but had an Egyptian way of life. He knew of pastoral nomads living along the Libyan coast.

The Persians also campaigned to the north of the Black Sea, in the land of Scythia, and Herodotus visited that country and described its peoples and culture in Book IV (Melpomene), and in the same book he wrote of peoples living in the regions beyond Scythia. He spent time in Athens and in 443 B.C.E. was present at the founding of a Greek colony at Thurii in southern Italy. It is likely that Herodotus spent the remainder of his life at Thurii, working on his history of the wars. He often referred to himself as Herodotus of Thurii.

Herodotus was born in Halicarnassus, a Greek city on the coast of what is now Turkey. At that time the district that included Halicarnassus was ruled by Artemisia, a queen who owed allegiance to the Persians. After her death, Artemisia was succeeded by her grandson Lygdamis, who became a tyrant. Herodotus either left Halicarnassus or was exiled from it and went to live on the island of Samos, in the Aegean Sea. He later returned to Halicarnassus at about the time Lygdamis was overthrown. He visited Athens, where he may have given public readings from his history, and become friendly with the dramatist Sophocles (ca. 496–ca. 406 B.C.E.).

PYTHEAS AND HIS VOYAGE TO THULE

Pytheas was a Greek explorer who traveled widely (see "Tin from Cornwall, Ivory and Peacocks from Asia" on pages 66–70). He visited Britain, and in *On the Ocean,* a work that was later lost, he also described a strange land he called Thule. He wrote *On the Ocean* some time between 330 B.C.E. and 320 B.C.E.; all that modern historians know of its contents is taken from ancient historians who saw it centuries later, principally Strabo (63 or 64 B.C.E.–ca. 24 C.E.), Pliny the Elder (23–79 C.E.), and Diodorus Siculus (first century B.C.E.).

Pytheas wrote that Thule was an island to the north of Britain that took six days to reach. One day's sailing northward from Thule, he reported that the "congealed sea" began and in summer the Sun never set. Thule was an agricultural country, he wrote, and produced cereals, fruit, milk, and honey. Unlike southern Europeans, who threshed their grain outdoors, the inhabitants of Thule had barns in which they threshed their grain—probably oats—and they made mead from their honey. When he was there the night lasted for only two or three hours.

Pytheas wrote that around Thule neither land nor sea nor air existed, but there was a mixture of all three, which he described as resembling a "marine lung"—the Greek name for jellyfish. That sounds like a description of *grease ice*—a covering of ice crystals that makes the sea appear oily and that is starting to separate into flat pieces of ice.

No one knows what land Pytheas had seen. Some scholars have suggested Thule may have been the Scottish Northern Isles—the Orkney or Shetland islands—but the southernmost Orkney Islands are close to the northern coast of mainland Scotland, and even the Shetland Islands, which are much farther away, are not six days' sailing from the mainland. In any case, neither group is far enough to the north to be surrounded by winter sea ice. Others have suggested Thule might be the Faeroe Islands, Iceland, or, perhaps, Greenland.

Alternatively, Pytheas's ship may have drifted eastward, carrying him toward Norway or to islands in the Baltic Sea. According to Pytheas, the journey to Thule commenced from the island of Berrice, which he described as being the largest of all. This suggestion of a group of islands may be a reference to the Outer Hebrides, off western Scotland—in which case Berrice, the largest, would be the Isle of Lewis. If Pytheas had sailed from there around the north of Scotland, continuing across the North Sea would have brought him to Norway, in the vicinity of Trondheim.

XENOPHON AND THE TEN THOUSAND

In 404 B.C.E. Artaxerxes (ca. 435 or 445–358 B.C.E.) became ruler of the Persian Empire. The son of Darius II, who had ruled from 423 B.C.E., Artaxerxes had a younger brother, Cyrus the Younger (424–401 B.C.E.), who sought to overthrow him and seize the Persian throne. Cyrus had close relations with the Spartans, with whom the Persians were allied in their war against Athens, and Greek troops formed the core of the forces he united into an army to fight his brother. His troops were mercenaries. They included archers, but there were 10,400 *hoplites* (citizen-soldiers who fought with a spear and round shield) and 2,500 *peltasts* (light infantry armed with javelins). The army came to be known as the Ten Thousand. In addition, Cyrus had at his disposal 100,000 Persian troops (although modern historians believe there were no more than 20,000) and 60 triremes (see "Biremes and Triremes" on pages 11–13).

In 401 B.C.E. the Ten Thousand marched from Greece all the way to Cunaxa, in Babylonia (modern Iraq), where they fought Artaxerxes. They won the battle, but because Cyrus was killed during the fighting, the victory was pointless—Artaxerxes remained ruler of Persia. At first Cyrus had concealed the true purpose of the campaign, and when the Greeks learned they were to fight the Persian king, the troops had refused to fight. Clearchus, the Spartan general in command of them, persuaded his men to fight, but after the battle Clearchus, along with four other generals and a number of other senior officers, was invited to peace talks, where they were all killed on the orders of Tissaphernes, the Persian regional governor. Now leaderless, and marooned deep inside enemy territory, the Ten Thousand sought an escape route.

Xenophon (ca. 444–357 B.C.E.) had accompanied the army. He was a professional soldier—that is, he was a mercenary, who fought for the general who was most likely to pay his wages—but he was also highly educated and intelligent and had been a pupil of the philosopher Socrates (ca. 469–399 B.C.E.). Xenophon marched with the Ten Thousand and later he wrote the story of their epic journey in a book called *Anabasis,* which means "the march up-country."

With Cyrus and Clearchus both dead, the Greek soldiers elected new generals—one of them was Xenophon—and began fighting their way northward, following a route parallel to the River Tigris that took them through Assyria and Armenia, over mountains, and finally to Trapezus (modern Trabzon) on the Black Sea coast. When, from the top of a mountain, they caught their first glimpse of the sea with its promise of Greek colonies along the coast, Xenophon reported that the Greeks gave one of the most famous shouts of joy in all history. "Thalatta, thalatta!" they cried ("the sea, the sea"). From there they headed westward, but when they reached Thrace, to the south of the River Danube, they found themselves fighting on the side of its ruler, and that conflict ended with their conscription into a Spartan army led by the general Thibron.

Anabasis was a literary success throughout Greece and Macedonia. Written in a very direct, straightforward style, it tells a stirring adventure story. It also shed light on Greek attitudes, because as they marched out of Persian territory the soldiers had to live off the land, decide how to deal with skirmishing bands, and determine the best route to follow. In Xenophon's account they made their decisions by debating the options and anyone could contribute ideas. It was a democratic process and the Ten Thousand are sometimes referred to as a "marching republic."

Xenophon was born near Athens into an aristocratic family. After his military adventures he was exiled from Athens. This may have been because he supported Socrates, who was then under suspicion of impiety and of corrupting the young, but it is more likely to have been because he had fought in a Spartan army against the Athenians. He went to live at Scillus in Elis, not far from Olympia in southern Greece, in a region controlled by Sparta, where he devoted himself to writing, hunting, and entertaining. In addition to *Anabasis,* Xenophon wrote treatises on hunting and on the horse.

After about 20 years, in 371 B.C.E. the Thebans defeated the Spartans at the Battle of Leuctra. This ended Spartan supremacy in Elis and the local people expelled Xenophon from Scillus. His exile from Athens had been revoked, but there is no evidence that Xenophon ever returned there. Historians believe he retired to Corinth, where he remained until his death in about 357 B.C.E.

ROMAN ROAD MAPS

The Romans once ruled an empire that stretched from the Atlantic to the southern shores of the Black Sea. It included what are now England and Wales in the far west, Turkey, Greece, Israel, and Lebanon in the east, as well as the whole of Egypt and a broad strip of North Africa. The authorities in Rome appointed governors to each region, with civil servants administering tax collection, trade, and law enforcement, and troops (many of them recruited locally) maintaining law and order as well as guarding the frontiers. Soldiers and officials were constantly on the move, as were traders. The Romans built roads to facilitate military movements, and others made use of the military roads. The Romans and those who worked for them were great travelers.

Travelers find maps very useful. Maps also allow conquering emperors to display pictorial representations of the lands they have seized. Ever practical, the Romans managed to combine the two approaches. They compiled a map of the empire that was also a schematic road map for travelers.

It was Julius Caesar (100–44 B.C.E.) who first dreamt of such a map. He engaged four Greek geographers to produce the map for him, and work began in the year he was assassinated. All work on the map ceased with the emperor's death. The idea of a map of the empire did not die with Caesar, however. At the time of Julius Caesar's death the man he had designated as his heir, Gaius Octavius (63 B.C.E.–14 C.E.) was studying in Illyria (now Albania). When word reached him, Octavius hastened back to Rome and became emperor. He was 18 years old and at first his position was weak, but he proved a skilled politician and in 30 B.C.E., after his armies had gained control of Egypt and his rivals Marcus Antonius (ca. 82–30 B.C.E.) and Cleopatra (69–30 B.C.E.) had committed suicide, Octavius became emperor in fact as well as in name. In 27 B.C.E. the senate conferred on him the imperial title *Augustus.*

As the emperor Augustus, he revived the idea of the map and handed the task to his most able and trusted general, Marcus Vipsanius Agrippa (63–12 B.C.E.). The two had been friends since boyhood and had served together as officers in the Roman cavalry. Agrippa helped the new emperor institute the reforms he believed desirable. In addition to being a very able general and admiral, Agrippa was keenly interested in architecture, sponsored art exhibitions, and was active in public life. He was also a geographer and the imperial map was drawn under his direct supervision. The work took a team of surveyors almost 20 years and Agrippa died before it could be completed. Augustus then took over the supervision. When the map was finished, the emperor ordered it to be carved in marble and erected on the wall of the Porticus Vipsaniae, a colonnade named in honor of Agrippa's sister. The carved map was lost when the Porticus disappeared and the master map on which it was based did not survive, but many copies were made of the map. These were drawn on papyrus scrolls and distributed to administrators and military commanders throughout the empire. None of those copies has survived, but later copies were made and one of those, drawn in the fourth century, did survive. It was copied again by a monk in the 13th century. Then the 13th-century copy was lost until Conrad Celtes (1459–1508), a German scholar, rediscovered it in a library in Worms, Germany. Celtes intended to publish the map, but died before he was able to do so; he bequeathed it to a friend who was an antiquarian, Konrad Peutinger (1465–1547). The Peutinger family kept it until 1714, and in 1737 the Austrian emperor purchased it and placed it in the Imperial Court Library. The map is called the Tabula Peutingeriana, or Peutinger Table, and it is now held in the Österreichische Nationalbibliothek (Austrian National Library), in Vienna.

The Peutinger Table is a parchment scroll 13 inches (0.34 m) wide and 22 feet (6.7 m) long and designed to be folded into a portfolio. It is made from 11 sections, but a 12th section, showing the British Isles and Iberian Peninsula, is missing. The illustration on page 114 shows a part of it.

The original map covered the area from the southeastern tip of England and the Pyrenees to India, Sri Lanka (called Insula Trapobane), and China and the coast of the imagined eastern ocean in the east, as well as North Africa. It had 534 illustrations, but it greatly distorted shapes. The Nile appeared to flow east–west, and the Mediterranean and Black Seas were long and narrow, looking rather like canals. The map showed more than 50,000 miles (80,000 km) of paved roads, with distances in Roman miles (1 Roman mile = 0.944 mile = 1.52 km) indicated by marking the positions of milestones. It also showed provincial boundaries, rivers, towns, and cities, and

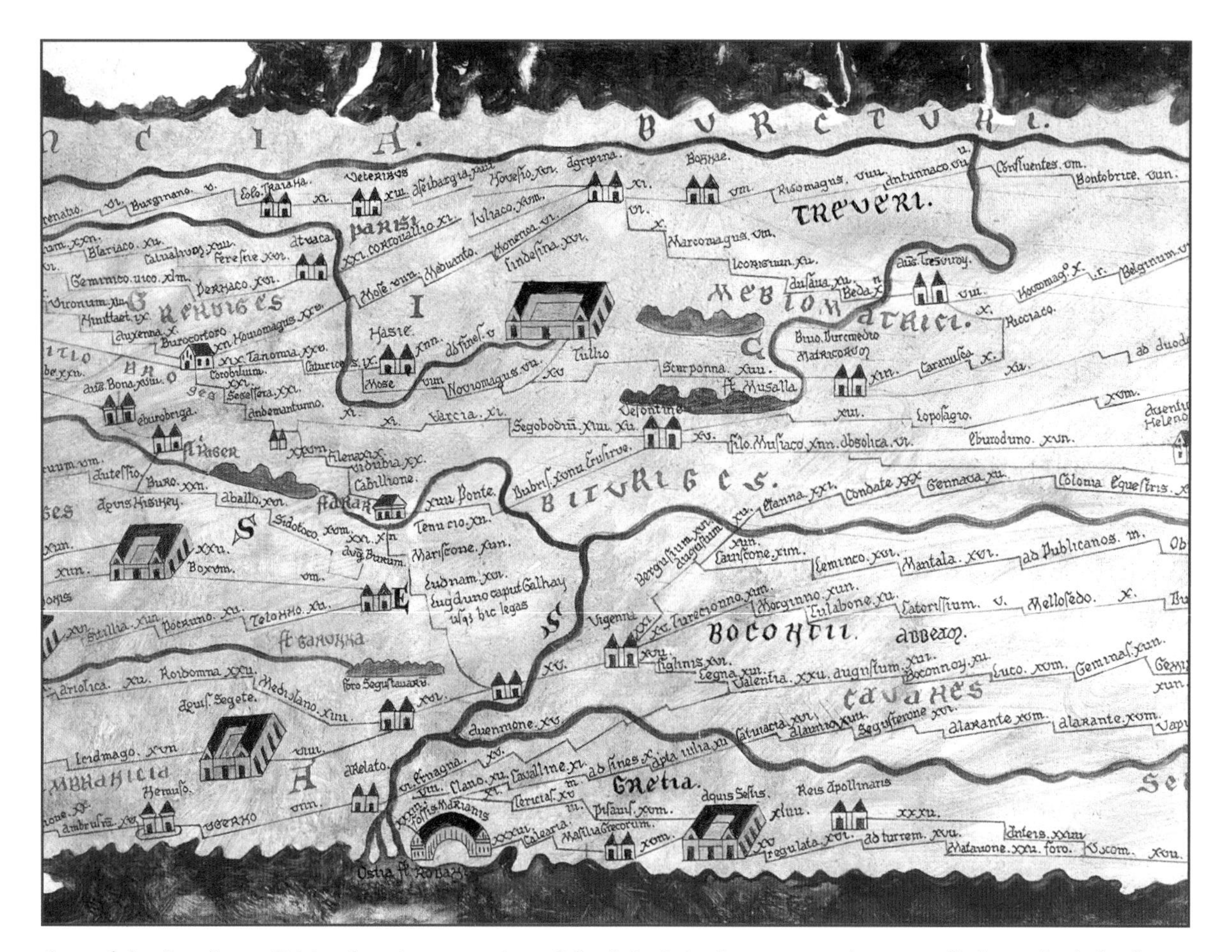

Part of the Peutinger Table, showing a section of Gaul; Paris is shown near the top, a little to the left of center. This is a 13th-century copy of a fourth-century copy of a map compiled three centuries earlier. It shows rivers, roads, distances, temples, baths, and other important buildings. *(Archives Charmet)*

facilities for travelers such as baths and temples. It was not drawn to scale, but as a schematic map it would have helped travelers to plan their journeys and find their way.

IBN BATTUTAH, THE GREATEST OF ALL MUSLIM TRAVELERS

Rihlah is an Arabic word that means journey, and is also the short name for a work with the English title *A Gift to Those Who Contemplate the Wonders of Cities and the Marvels of Traveling.* The *Rihlah* is the record of a journey of approximately 75,000 miles (120,675 km) that took its author, Ibn Battutah (1304–68 or 1369), to almost all of the Islamic countries, as well as to China and as far as Sumatra. The journey lasted for 24 years and the resulting book is one of the greatest of all pieces of travel writing.

Ibn Battutah's full name was Abu 'abd Allah Muhammad Ibn 'abd Allah Al-lawati At-tanji ibn Battutah; he was born on February 24, 1304, in Tangier, Morocco. He came from a family of lawyers and studied law and literature in Tangier. His travels began in 1325, with the *hajj,* the pilgrimage to Mecca. It took Ibn Battutah 16 months to reach Mecca, traveling across North Africa and staying in Tunis for two months. Whenever possible, he chose to travel by caravan for safety (see "Caravans and Oases" on pages 52–56).

By the time he reached Egypt on his way to Mecca he had been seized by the urge to continue traveling and he resolved never to travel along any road for a second time. Others traveled as refugees or pilgrims, or to seek a better life, as soldiers marching to a distant war, or as merchants, but Ibn Battutah traveled simply for the joy of being on the move and experiencing novel sights, sounds, and smells. He arrived at Alexandria in 1326 and spent several weeks there before heading inland to Cairo, where he stayed for about one month. There were several routes from Egypt to Mecca. Ibn Battutah chose one of the less popular ones, planning first to head up the Nile, then eastward to the Red Sea port of Aydhab, and from there by sea to Arabia—but as he approached Aydhab there was a local rebellion and the way was blocked. So he returned to Cairo and went from there to Damascus, Syria, where he spent the month of Ramadan before joining a caravan heading for Medina and Mecca.

His plan was to complete the pilgrimage but then, instead of returning home, continue his education by studying under prominent scholars in Syria, Egypt, and Hejaz, the region of western Arabia that contains the cities of Mecca and Medina. He had qualified as a lawyer in Damascus, and the other diplomas he received marked him as an educated man.

He spent one month in Mecca and on November 17, 1326, he joined a caravan of pilgrims returning across the desert to their homes in Mesopotamia. Ibn Battutah arrived in Najaf, south of Baghdad, and then visited southern Persia and Azerbaijan before he returned to Mesopotamia, reaching Baghdad in June 1327. That part of the world was then ruled by Mongol khans, and Ibn Battutah met Abu Sa'id (1305–35), the last khan of Persia. Abu Sa'id was leaving the city and Ibn Battutah traveled northward with his caravan through Mosul and into what is now Turkey. He then returned to Baghdad, where he joined a caravan heading back to Mecca.

Ibn Battutah made a second hajj and then spent some time in Mecca and Medina before leaving from the port of Jeddah in 1328 or 1330 and arriving at Yemen. He crossed Yemen by land to the port of Aden, from where he sailed along the African coast, visiting Mogadishu, Mombasa, Zanzibar, Kilwa, and other cities, returning to Mecca by way of southern Arabia, Oman, Hormuz, southern Persia, and the Persian Gulf. He arrived back in Mecca in 1330 or 1332.

He spent another year in Mecca, and then decided on a much more ambitious journey, this time to visit Muhammad ibn Tughluq (ca. 1290–1351), the Sultan of Delhi. He began by heading northward, visiting Egypt and Syria, and arriving by sea in Anatolia (modern Turkey), where his travels took him to many of the small sultanates into which Anatolia was then divided. From Anatolia he crossed the Black Sea to the Crimea, passed through the northern Caucasus, and reached the Volga River and Saray, the capital of Muhammad Ozbeg (1282–1341), khan of the Golden Horde. The khan's wife, a Byzantine princess, was pregnant and wished to return to Constantinople for the birth. The khan arranged a party to accompany her and Ibn Battutah managed to join it. This journey took him outside the Muslim world for the first time. In 1332 he met the emperor of Byzantium, Andronikos III Palaiologos (1297–1341). He spent a month in Constantinople, and then crossed the steppes eastward in the direction of India, passing through Bukhara and Samarkand, crossing Afghanistan, and eventually he arrived at the Indus River.

By this time Ibn Battutah was a man of importance with a reputation that preceded him, and a large retinue of wives, concubines, servants, and followers accompanied him. Muhammad ibn Tughluq received him warmly and after a time appointed him a *qadi* (a judge administering Islamic law), an official position carrying few duties that Ibn Battutah held for several years. It soon transpired that life in India would not be without risk, however. The sultan's behavior was erratic and he could be extremely cruel. Ibn Battutah came to fear for his life as a number of his friends fell foul of the tyrant and were killed. Eventually, Ibn Battutah fell under suspicion, but then the Sultan's mood changed and in 1342 he despatched Ibn Battutah as his envoy to the emperor of China. Ibn Battutah had to cross India and as he did so he was waylaid by insurgents, became caught up in local wars as he neared the coast, and when at last he set sail his ship was wrecked near Calicut (Kozhikode), losing the sultan's gifts for the emperor. Fearful of the sultan's wrath, he spent some time in southern India and then sailed to the Maldives, where he spent nearly two years. As a *qadi* he became active in politics, married a member of the ruling family, and even thought of becoming sultan. His popularity waned, however, when he grew critical of what he regarded as the lax morals of the islanders and he decided it was time to leave. He sailed to Sri Lanka. When he left Sri Lanka his ship was wrecked off the Indian coast and pirates attacked the ship that rescued him. Back in India, Ibn Battutah took part in a war led by one of his brothers-in-law, then found his way back to Calicut, from where he sailed back to the Maldives, visited Bengal and Assam, and then sailed to Sumatra, where the sultan provided him with a ship to take him to China.

At last Ibn Battutah arrived in China, at the port of Hangzhou, in Fujian Province. From there he claimed to have sailed along rivers and canals as far as Peking (Beijing), then returned to Hangzhou, although there is some doubt over the veracity of this part of his story. He left Hangzhou and traveled by Sumatra, Malabar, and the Persian Gulf to Syria and Baghdad. He reached Damascus in 1348, during the Black Death. He visited many towns in Syria and Egypt, made a final pilgrimage to Mecca, and sailed from Alexandria to Tunisia, from there to Sardinia and Algiers, and finally arrived in Tangier in November 1349. He had learned of his father's death in Damascus and when he reached home he discovered his mother had also died.

Soon after returning to Tangier, Ibn Battutah left again, this time to visit Al-Andalus, in southern Spain. In 1352 he visited western Sudan. His final journey was across the Sahara to West Africa, spending a year in the Empire of Mali.

Towards the end of 1353 Ibn Battutah began the journey home in a large caravan transporting female slaves. He arrived in Morocco early in 1354. At the request of the Sultan of Morocco, Ibn Battutah dictated the story of his travels to Ibn Juzayy (1321–57), a scholar he had met in Spain, and Ibn Juzayy transformed Ibn Buttatah's simple account into a work of literature. Little is known of Ibn Battutah's final years. He was a *qadi* in a town in Morocco and he died in 1368 or 1369. He was buried in Tangier.

FRIAR ODORIC AND HIS JOURNEY TO INDIA, CHINA, AND TIBET

In 1245, Pope Innocent IV sent an emissary to the Mongol ruler Güyük Khan (ca. 1206–48) with a message pleading for him to convert to Christianity and end his attacks in Europe. The khan replied, calling on the pope and other European rulers to submit to him. Pope Innocent persisted, sending more missions, all with the aim of winning converts and bringing an end to Mongol hostility. The pope entrusted many of these missions to Franciscan friars.

Odorico Mattiussi (ca. 1286–1331) was one such friar. He was born near the town of Pordenone, in Friuli, Italy, joined the Franciscan order at an early age, and took his vows at Udine, the capital of Friuli. Later he became known as Odoric of Pordenone. It was in about 1316 or 1318 that he was despatched eastward as a missionary.

Odoric departed from Venice and visited Constantinople, sailing from there to Trebizond (modern Trabzon, Turkey) and Erzurum, where he stayed in Franciscan houses, and then traveling overland through Anatolia and into Persia (Iran). He visited Persepolis, once the Persian capital, crossed Mesopotamia (Iraq) with a visit to Baghdad, and reached the Persian Gulf. He sailed from Hormuz to India, landing at Thana, near Mumbai, in about 1322, where four members of his order had recently been put to death on the order of the Muslim governor. Odoric disinterred their remains, which had been buried to the north of the city, and carried them with him. He visited various parts of India and may have visited Sri Lanka before he embarked on a Chinese junk bound for Sumatra. He explored the northern coastal regions of Sumatra and from there he sailed to Java and possibly to Borneo before reaching Indochina, from where he sailed to China.

Odoric spent some time in Guangzhou and other coastal cities. He founded Franciscan houses at Zayton (Quanzhou) and Xiamen and left at one of them the remains of the martyrs he had carried from Mumbai. He reached Fuzhou and then headed inland, visiting Hangzhou, then held to be the greatest city in the world and which he described in detail. He passed through Nanjing and sailed along the Grand Canal for the last stretch of his journey to Beijing, where he remained for three years, possibly from 1324 to 1327, staying in a Franciscan house.

There is good evidence for the authenticity of Odoric's account of his travels. He reported seeing Chinese fishermen using tethered cormorants to catch fish, mentioned the fashion for growing fingernails to extraordinary lengths, and he also described the binding of women's feet. No earlier traveler had mentioned these things.

After leaving Beijing, Odoric headed for home, traveling westward. He probably crossed Mongolia and Tibet, visiting Lhasa, before entering northern Persia. He reached Venice in 1329. In the course of his long journey, Odoric baptized more than 20,000 persons.

After his return to Italy Odoric moved to a Franciscan house in Padua where, in 1330, Friar William of Solagna wrote an account of Odoric's travels in Latin. The written story of Odoric's travels was first published as a book in 1513. The English writer Richard Hakluyt (1552 or 1553–1616) included them in his *The Principal Navigations,* published in 1589, and they have appeared in several other editions, some in French and others in Italian.

Later in 1330 Odoric set off on a journey to the papal court at Avignon, hoping to present himself to Pope John XXII, but he fell ill. He turned back to Udine, where he died on January 14, 1331. By the time of his death Odoric was a popular figure, but he had made much less of an impression on his order. The Franciscans were about to bury him in a simple ceremony when the town's chief magistrate intervened and ordered a public funeral. News spread and the funeral had to be postponed several times. Odoric became a figure of popular devotion, with miracles attributed to him, and the municipality erected a shrine to hold his body. In 1775, Odoric was beatified by Pope Benedict XIV.

PRINCE HENRY THE NAVIGATOR AND THE AFRICAN COAST

In about 1165, at a time of great tension between the Christian and Muslim worlds, a letter addressed to Manuel I Comnenus (1118–80), the Byzantine emperor, reached Pope Alexander III (ca. 1100 or 1105–81), Frederick I Barbarossa (1122–90), the Holy Roman Emperor, and several other European rulers. The letter, written in Latin, purported to be from a Christian priest calling himself Presbyter Johannes, which became corrupted to Prester John. Prester John said he lived as a simple priest but ruled an empire to the east of the Islamic lands, where peace reigned, everyone enjoyed unimaginable wealth, and his court was staffed by archbishops and kings. Prester John said he had heard of good, Christian rulers in the west, that he wished to make contact with them, and that, with their agreement, he intended to send his vast armies westward to regain the Holy Sepulchre. The legend of Prester John had ancient origins, but the letter was believed. The pope replied to it and several expeditions headed east in the hope of making contact with this fabled ruler. The letter was later revealed to be a hoax. There was, indeed, a powerful ruler far to the east, but far from being a simple Christian presbyter he was Genghis Khan (ca. 1162–1227), the Mongol warlord. Nevertheless, the dream inspired many travelers. One of them was Prince Henry (1394–1460), called "the Navigator" because of the many voyages he sponsored, although he sailed on none of them.

Henry was born in Oporto, Portugal, on March 4, 1394, the third son of John I (1357–1433), king of Portugal, and Philippa of Lancaster (1359–1415), the eldest daughter of John of Gaunt. In those days pirates, also called corsairs, were harassing Portuguese coastal communities, capturing villagers and selling them as slaves in Africa. The pirates operated from the Barbary coast, which extended from Libya westward to the Atlantic. In 1415 Henry and his brothers persuaded their father to mount an attack on Ceuta, one of the corsairs' ports on the North African side of the Strait of Gibraltar. They sailed in July and their attack succeeded. Ceuta was undefended, the Portuguese captured it, and the king immediately appointed Henry its governor, a position that required him to ensure the port was adequately defended, but not to live there. When Muslim forces tried to retake the city in 1418, the Portuguese garrison fought them off before Henry could arrive with reinforcements.

His exposure to Africa aroused Henry's interest in exploring more of the coast. He hoped to discover the gold of Africa and, eventually, to find a sea route to India and, perhaps, to make contact with Prester John. He began to sponsor short exploratory voyages in 1418 and in 1419, when, as duke of Viseu and lord of Covilhã, he was appointed governor of the Algarve, in southern Portugal, he founded his own court at Sagres, on the southwestern tip of the Iberian Peninsula.

He attracted mariners, navigators, mapmakers, astronomers, and shipbuilders to his court and began to plan longer voyages. On May 25, 1420, Henry was appointed grand master of the Order of Christ, the Portuguese successor to the Knights Templar and sponsored by the pope. This gave him access to considerable funds that he used to finance his expeditions. Because it was his duty to convert pagans to Christianity, all his ships had a red cross on their sails.

The explorers needed a new type of ship that was lighter, faster, and more maneuverable than the vessels then plying the Mediterranean. Henry's shipbuilders designed and constructed the caravel. This became the preferred Portuguese (and later, Spanish) ship for long voyages.

The long voyages began in 1420, from the Portuguese port of Lagos (not to be confused with Lagos, Nigeria). Expeditions usually consisted of one or two vessels that followed the coast and spent each night moored close to the shore. In 1434 Gil Eanes, one of Henry's captains (about whom little is known), sailed past Cape Bojador, a headland on the coast of Western Sahara, becoming the first European to do so, although Hanno the Navigator may have done so 2,000 years earlier, sailing from Carthage (see "Pilot Books" on pages 88–90). The tides and currents around the headland are treacherous; until that time seafarers had believed the place was close to the end of the world, and they kept away from it.

In the same year ships commanded by João Gonçalves Zarco (ca. 1390–1471), Bartolomeu Perestrello (ca. 1395–1457), and Tristão Vaz Teixeira (ca. 1395–1480) reached the Madeira Islands and the Portuguese colonized them. A monk, Frei Gonçalo Velho (dates of birth and death unknown), commanded the ship that reached the Azores in 1427. The Portuguese later colonized these islands, also. Other expeditions pressed farther along the African coast. They reached Cape Blanco, Mauritania, in 1441 and the Bay of Arguin, some distance to the south, in 1443. In 1445 they reached the mouth of the Senegal River and soon after that they sighted the Gambia River. They had then achieved one of Henry's ambitions by reaching lands to the south of the desert that were not under Muslim control. The first cargo of African gold and slaves arrived in Portugal in 1441, silencing critics who had been complaining that Henry was wasting money on fruitless voyages. By 1448 the first overseas European trading post was established on Arguin Island; the trade was in slaves. By

1456 Henry's ships had reached the Cape Verde Islands and by 1462 they had reached Sierra Leone. That is as far as they traveled, and the expeditions left Henry in debt.

In 1433 King John died and Henry's brother Duarte succeeded him and immediately began criticizing Henry for the money he was spending. Duarte reigned for five years. In 1437, a year before Duarte died, Henry and Fernando, the third brother, obtained his permission to attack Tangier. The enterprise was disastrous and Fernando was captured and held as a hostage, dying from ill treatment in 1443. After Duarte's death in 1438, there were family quarrels that led to war. Henry tried unsuccessfully to act as peacemaker. He died at Sagres on November 13, 1460.

In 1459, the year before he died, Fra Mauro, an Italian monk from Murano, an island near Venice, completed a map of the world

Fra Mauro completed his map of the world in April 1459. The original has not survived, but shortly before his death Fra Mauro began to make a copy for the Venetian authorities. He died before completing it, but his colleague Andrea Bianco finished it. This is that copy, now held at the Biblioteca Nazionale Marciana, Venice. *(Scala/Art Resource, NY)*

that had been commissioned by Henry's nephew, King Afonso V (1432–81). Fra Mauro had consulted sailors, sometimes paying them for information, and his map was by far better than any earlier map, most of which depicted a schematic view of the world based on Christian theology. Fra Mauro aimed to show the world as it really was. The map was sent to Portugal by the ruler of Venice, accompanied by a letter addressed to Prince Henry, urging him to continue funding voyages of exploration. The original map has not survived. The illustration on page 122 shows the copy Fra Mauro started to make and that a colleague completed after his death.

MARCO POLO AND HIS TRAVELS IN ASIA

The Travels, Marco Polo's account of his adventures, was written in about 1298 and proved so popular that it has never been out of print. As a result, Marco Polo is probably the most famous traveler in history. The book was written while Polo was a prisoner-of-war of the Genoese, captured during a war between Genoa and Venice. While he was imprisoned, Marco described his travels to a fellow prisoner, Rustichello da Pisa, a writer of romances, and they wrote *The Travels* together. The first edition was written in Old French and entitled *Le divisament dou monde* (The description of the world). That version is lost, but it had already been translated into Latin and from Latin into Italian. The book was soon translated into several other languages.

Marco Polo was born in Venice on September 15, 1254. His father, Niccoló Polo, was a merchant. Niccoló and his two brothers Maffeo and Marco were business partners trading with Asia. In 1259, Niccoló and Maffeo sailed from Constantinople (now Istanbul), where they had been living for several years, to Sudak, a port in the Crimea where there was a Venetian colony and where their brother Marco owned a house. After a time they left Sudak, forming or joining a caravan transporting goods to Sarai Batu, a city on the River Volga where Berke Khan, ruler of the Mongol state known as the Golden Horde, had established his capital. They remained there for about a year, but when they wished to return they found their route blocked by a war that had broken out between Berke and his cousin Hulagu Khan. The Polo brothers sought refuge in Bukhara (in modern Uzbekistan), which was neutral. Stranded in Bukhara, the brothers learned the Mongol language and when they were offered an opportunity to join a diplomatic mission from Hulagu to his brother, Kublai Khan, they gladly accepted.

In 1266 the Polo brothers reached the city of Dadu (modern Beijing), Kublai's capital. Kublai, an intelligent, humanitarian, highly cultured man, welcomed the two travelers warmly and listened to all they had to tell him about the West. Kublai sent them back to Europe with a safe conduct—a kind of passport—as well as gifts for the Pope and a letter requesting 100 men to teach his people about Christianity and western customs and some holy oil from Jerusalem. The Polos reached Italy in 1269.

Pope Clement IV had died in 1268 and it was 1271 before Gregory X was elected to succeed him. Pope Gregory received Kublai's letter and gifts and blessed the Polos. He could not find 100 scholars, but he sent the oil and two Dominican monks. The monks gave up, but the two Polo brothers, this time accompanied by the 17-year-old Marco, persevered and in 1274 they reached Dadu and presented Kublai with the gifts sent by the Pope. Marco was an excellent linguist and entertaining storyteller, and he became a favorite with Kublai Khan.

The three Polos remained at the court of Kublai Khan for 17 years, during which time the Khan sent Marco on various missions around the country. In 1291 Kublai released them and they set out for home, escorting a Mongol princess who was to be the bride of Arghun Khan, the Mongol ruler of the Levant. The journey took them two years, sailing through the Malay Strait, around India to Hormuz at the southern end of the Persian Gulf, overland to Trebizond (in modern Turkey), and by sea from there to Venice.

The war between Venice and Genoa ended in 1299 and Marco was released. He returned to Venice, where his family had bought a large house out of the profits from their travels. Marco was now wealthy. In 1300 he married Donata Badoer and they had three daughters, Fantina, Belela, and Moreta, all of whom later married into aristocratic families. Marco died at home in January 1324, leaving most of his possessions to be divided between his children.

JOHN CABOT AND THE DISCOVERY OF NORTH AMERICA

Christopher Columbus crossed the Atlantic in 1492 on behalf of Spain (see "The Voyages of Christopher Columbus" on pages 147–152). This alerted the English to the possibilities of lands and riches far to the west, as well as to the danger that the Spanish and Portuguese might claim them. On May 4, 1493, Alexander VI issued a papal bull granting to Spain all the lands to the west and south of a line 260 miles (418 km) west of the Azores or Cape Verde Islands and to Portugal all the lands to their east and north. Henry the Navigator had sponsored exploratory voyages seeking a sea route around Africa to Asia, and Columbus had sailed westward in the hope of reaching Asia by that route.

On March 5, 1496, the English King Henry VII (1457–1509) issued letters patent to an Italian navigator living in Bristol. The term *letters patent* describes a legal document issued by a monarch or government in the form of an open letter granting certain rights or status to an individual or entity such as a city or corporation. In this case, the document included the following:

> Be it knowen, that We have given and granted, and by these presentes do give and grant for Us and Our Heyres, to our well beloved John Cabote, citizen of Venice, to Lewes, Sebastian, and Santius, sonnes of the sayde John, and to the heires of them and every of them, and their deputies, full and free authoritie, leave, and Power, to sayle to all Partes, Countreys, and Seas, of the East, of the West, and of the North, under our banners and ensignes, with five shippes, of what burden or quantitie soever they bee: and as many mariners or men as they will have with them in the saide shippes, upon their owne proper costes and charges, to seeke out, discover, and finde, whatsoever Iles, Countreyes, Regions, or Provinces, of the Heathennes and Infidelles, whatsoever they bee, and in what part of the worlde soever they bee, whiche before this time have been unknowen to all Christians. We have granted to them also, and to every of them, the heires of them, and every of them, and their deputies, and have given them licence to set up Our banners and ensignes in every village, towne, castel, yle, or maine lande, of them newly founde. And that the aforesaide John and his sonnes, or their heires and assignes, may subdue, occupie, and possesse, all such townes, cities, castels, and yles, of them founde, which they can subdue, occupie, and possesse, as our vassailes and lieutenantes, getting unto Us the rule, title, and jurisdiction of the same villages, townes, castels, and firme lande so founde.

The letters patent ended "Witnesse our Selfe at Westminster, the Fifth day of March, in the XI yeere of our reigne. HENRY R".

John Cabot and his sons were authorized to search for previously unknown lands and to bring any merchandise they acquired in such lands to the port of Bristol. The document also granted them a monopoly on all trade they were able to establish in the lands they discovered. They would not pay customs duty on goods they brought to England, but would pay one-fifth of the profits to the crown. The document exhorted all subjects of the crown to assist the Cabots in every way. Enough Bristol merchants were attracted by the trade monopoly to invest in the enterprise.

John Cabot was the English name of Giovanni Caboto (ca. 1450–ca. 99). It is not certain where he was born, but in 1461 Cabot moved to Venice and he became a Venetian citizen in 1476. He worked for a Venetian merchant, traveling to the eastern Mediterranean and visiting Mecca. He gained experience as a sea captain and became a skilled navigator.

Like Columbus, Cabot believed it possible to reach Asia by sailing westward, but unlike Columbus he planned to take a northerly route to pass around America to the north. His reasoning was that the meridians of longitude are closer together in the far north, so the distance should be shorter. Silks, gems, and spices were the prizes tempting both men and their sponsors (see "Silks and Spices" on pages 57–59). These reached Europe by the overland route from Asia, incurring taxes and heavy carriage charges all along the way. If they could be transported by sea, in ships belonging to the importer, and then retailed at a price only a little lower than before, the profits would be vast. Cabot sought sponsorship in Spain and Portugal, but without success, because geographers in those countries did not think Cabot's northerly route existed. It is not certain when Cabot and his sons moved to England, but they were living in Bristol by the end of 1495, and Cabot sought support from the king, this time successfully.

Cabot embarked on his first voyage in 1496, with only one ship, but when he reached Iceland bad weather, a shortage of food, and disputes with his crew forced him to turn back. He made a second attempt in May 1497 on the *Matthew* (or possibly *Mattea,* the name of Cabot's wife), a three-masted caravel, 78 feet (24 m) long and 20.5 feet (6.2 m) wide, with a crew of 18. They sailed around the south of Ireland then headed northwest and made landfall in North America on the morning of June 24, either in southern Labrador, Cape Breton Island, or at Cape Bonavista or St. John's, Newfoundland. He and his crew believed they had reached the northern coast of Asia. They saw signs that the place was inhabited but met no people, and Cabot claimed the land for England, planting the English and Venetian flags and naming a number of geographic features. They set sail for home, probably on July 20, and took a more southerly course, arriving at Brittany by mistake but reaching Bristol on August 6. Cabot said that the land he had discovered enjoyed a temperate climate, the land was fertile, and the adjacent sea teemed with fish. He announced his intention to return and then continue westward until he reached Japan. He was made an admiral and given a reward of £10.

On February 3, 1498, he received new letters patent authorizing a new expedition. This time he took five ships and about 200 men. One of the ships sustained damage at sea, returned to Ireland for repairs, and then resumed its voyage, following the others. None of them was ever heard from again and although they may have reached America, most historians believe the entire fleet was lost at sea. In 1508 John's son, Sebastian Cabot (ca. 1484–ca. 1557) sailed to North America in search of the Northwest Passage (see "Franklin, McClure, and the Discovery of the Northwest Passage" on pages 159–161) and from 1525 to 1528 he explored the coast of South America searching for silver.

AMERIGO VESPUCCI IN SOUTH AMERICA AND THE CARIBBEAN

Both Columbus and Cabot believed that the lands they explored on the western side of the Atlantic were in Asia. These lands were called the Indies for some time. The first explorer to realize that this was not part of Asia but a New World was the Italian merchant and explorer Amerigo Vespucci (1451–1512). He came to this conclusion in the course of a voyage he made in 1501–02 and other scholars agreed with him. Of course, they all believed Columbus had been the first European to set foot in this New World—they did not know about the earlier Norse visit (see "Leif Erikson and Vinland" on pages 49–51)—and the New World should have been called Columbia, but it was not to be. Vespucci's published accounts of his travels were read far more widely than were those of Columbus. Vespucci included lively detailed descriptions of the way of life of the peoples he encountered that made his work popular. In 1507, the German cartographer Martin Waldseemüller (born ca. 1470, died between 1518 and 1521), who lived in Saint-Dié, in Lorraine, France, published a pamphlet with the title *Cosmographiae introductio* (Introduction to cosmography) followed by *Quattuor Americi navigationes* (Four voyages of Amerigo). In this, Waldseemüller proposed that the New World be named after Amerigo. He argued that the other continents—Europe, Asia, and Africa—had been named after women, so it was appropriate to give a female name to the new continent, and he adapted it from the baptismal name of the person who discovered that it was, indeed, a new continent. At about the same time Waldseemüller compiled a map of the world in which the name America was used for the first time, although it referred only to South America.

Amerigo Vespucci was born on March 9, 1451, at Florence, Italy, the son of a notary. He was educated by his uncle, Giorgio Antonio, a philosopher who tutored the children of many noble Florentine families. Amerigo's father died in about 1483, and Vespucci became a steward in the household of Lorenzo di Pierfrancesco de' Medici; from 1478 to 1480 he was attached to the Florentine embassy in Paris. On his return from France, Vespucci entered the Medici bank. In late 1491 the Medici sent him to Seville, Spain, where they owned a firm headed by Giannetto di Lorenzo Berardo Berardi, whose business consisted mainly in fitting out ships. Berardi helped Columbus prepare for some of his voyages. Vespucci collaborated in this and became friendly with Columbus. When Berardi died in late 1495 or early 1496, Vespucci took over the business as the Medici agent in Seville.

Vespucci obtained three ships from Ferdinand, king of Castile, and sailed as navigator from Cádiz on May 10, 1497, in search of a new route to Asia. He reached the South American coast in what are now Guyana or Brazil, and he may have entered the Gulf of Mexico and sailed along part of the North American coast, although this is uncertain. He arrived back in Cádiz on October 15, 1498.

His second voyage began on May 16, 1499, with a fleet of four ships commanded by Alonso de Ojeda (ca. 1465–ca. 1515). On reaching Guyana he turned south, discovering the mouth of the Amazon River, and he discovered the mouth of the Orinoco River on his way back. He visited Trinidad and Hispaniola (Haiti and the Dominican Republic), still thinking he was in Asia. He arrived back in Cádiz in June 1500, and soon began equipping a new expedition to find a western route to the Bay of Bengal and the island of Taprobane (Sri Lanka). The Spanish authorities would not finance this expedition, however, so Vespucci entered the service of the Portuguese government.

On May 13, 1501, Vespucci sailed from Lisbon to the Cape Verde Islands, then headed southwest, reaching the Brazilian coast at about 13° S on January 1, 1502, and reaching the bay where the city of Rio de Janeiro now lies. He claimed to have continued south along the Patagonian coast and perhaps sighting South Georgia, at 54° S, but since his account failed to mention passing the mouth of the Río de la Plata, this is doubtful. The expedition arrived back in Lisbon on July 22, 1502. Vespucci may have sailed on another expedition for the Portuguese in 1503–04, but this is uncertain. What is known is that he helped prepare later expeditions but did not take part in them.

Early in 1505 Vespucci was recalled to Spain and the Spanish government engaged him to work for the Casa de Contratación, the organization established in 1503 to supervise trade with Spanish colonies in America. He was made official navigator at the Casa in 1508. This was a very responsible position, requiring him to examine the licenses for ships and their masters planning voyages. He also supervised the preparation of the official map of America and the routes to it. Vespucci took Spanish citizenship and remained in this post until his death in Seville on February 22, 1512.

PEDRO ÁLVARES CABRAL AND THE DISCOVERY OF BRAZIL

Brazil is the world's fifth-largest country, with an area of 3.286 million square miles (8.512 km^2), yet Europeans discovered it almost by accident. In 1500, King Manuel I (1469–1521) of Portugal ordered a second expedition to India. Vasco da Gama (1460 or 1469–1524) had led the first, in 1497–98, but the king appointed Pedro Álvares Cabral (1467 or 1468–1520) to command the new fleet. Clearly, the king and his advisers believed Cabral had the navigational skill and experience to match the abilities of da Gama. Da Gama acted as a consultant, preparing the route Cabral would follow. Cabral was to strengthen commercial relations with India and conquer more territory.

Historians believe Cabral was born in Belmonte, the son of Fernão Cabral and Isabel de Gouveia. His father was a nobleman and his mother a descendant of King Alfonso I (1109–85), and Cabral's wife, Isabel de Castro, was also descended from Alfonso I. Manuel I regarded Cabral highly and bestowed several honors on him, including membership of the Order of Christ, which required Cabral to find converts to Christianity. The king made Cabral admiral of a fleet of 13 caravels. The captains accompanying him included Bartolomeu Dias (ca. 1451–1500), who in 1488 had become the first European known to have sailed round the Cape of Good Hope. Da Gama advised Cabral to follow the route he had pioneered, avoiding the Gulf of Guinea, the sea area to the south of West Africa, which was notorious for its long periods of calm. Instead, Da Gama said they should sail far to the west.

The fleet sailed from Lisbon on March 9, 1500, heading for the Cape Verde Islands. They encountered a storm after leaving the islands and one ship had to return to Lisbon. The remainder set a southwesterly course. This route had the added advantage of allowing Cabral and his colleagues to study the coast of the western lands that previous navigators had sighted from a distance. There had been a dispute between Spain and Portugal over the rights to land on the western side of the Atlantic. Pope Alexander VI had resolved this in 1493 with a papal bull (see "John Cabot and the Discovery of North America" on pages 124–127) and on June 7, 1494, Spain and Portugal had formalized an agreement, the Treaty of Tordesillas, under which Spain could claim all territory to the west of the line stipulated by the pope, Portugal could claim land to the east, but neither could claim land already claimed by a Christian ruler. The coastline that the Cabral expedition would study lay on the Portuguese side of the line.

On April 21, the fleet saw a mountain on the horizon and named it Monte Pascoal, and the following day a party landed on the shore of what they believed was an island. On April 25 the entire fleet sailed into the harbor at Porto Seguro—now a popular tourist destination. They named what they supposed was an island the Island of the True Cross *(Vera Cruz)* and erected an iron cross. King Manuel later changed the name to Island of the Holy Cross. Cabral took formal possession of the country and sent one of the ships back to Portugal to convey the news to the king. What the Cabral fleet had discovered was not an island, however, but a vast territory with ill-defined land borders that became a regular port of call for ships bound for the Cape of Good Hope and the Indian Ocean. Eventually it acquired its modern name (in Portuguese, Brasil; in English, Brazil), from a valuable dyewood found there (see the sidebar below).

The fleet sailed from Porto Seguro on May 3, having spent only 10 days there. They reached the Cape of Good Hope, but on May 29, as they were rounding the Cape, four of the ships sank in a storm, with no survivors. Bartolomeu Dias was among those lost. One ship, commanded by Diogo Dias, brother of Bartolomeu, later became separated from the rest of the fleet and on August 10 it discovered a large island the crew named St. Lawrence. They then returned to Portugal. The island was later renamed Madagascar.

What remained of the fleet reached Calicut, India, on September 13. The ruler allowed Cabral to establish a fortified trading post there

HOW BRAZIL ACQUIRED ITS NAME

In the 15th and 16th centuries wealthy Europeans dressed in bright colors, while ordinary folk were clad in dull grays and browns. Red was especially popular, not least because the dye, derived from the wood of the sappanwood tree *(Caesalpinia sappan),* was so expensive that sporting a red coat or gown was an ostentatious display of wealth. The dye was imported from southern Asia and reached Europe in powder form. The Portuguese called the sappanwood tree *pau-brasil, pau* meaning "wood" and *brasil* meaning "ember." In English, sappanwood was also known as *brazilwood.*

On April 22, 1500, Portuguese sailors reached the coast of South America. When they went ashore they saw trees that had dense, orange-red wood, and they soon discovered that the wood yielded a red dye. These pau-brasil trees *(Caesalpinia echinata),* also known as pernambuco and yielding a dye every bit as good as the dye from sappanwood, grew abundantly along the coast and beside river courses extending far inland. Later explorers sent samples back to Europe and within a few years the trees were being felled and shipped across the ocean in large amounts. It proved such a highly profitable industry that ships laden with pau-brasil timber were at risk from pirates. In 1555 a French expedition of two ships and 600 men led by Nicolas Durand de Villegaignon (1510–71), vice-admiral of Brittany, attempted to establish a French colony at Rio de Janeiro, which they called France Antarctique; they made this attempt partly in order to gain access to the priceless timber. The territory from which pau-brasil was gathered became known in Portuguese as Brasil and in other languages as Brazil.

Brazilwood is still used as the source of the dye brazilin and the timber is also used to make bows for stringed instruments. The brazilwood industry was very destructive, however, and it finally collapsed in the 18th century because so many of the trees had been felled that they were becoming rare. The species is now classed as endangered.

but, following disagreements with traders a large force attacked the post and killed most of the defenders before reinforcements could arrive. Cabral bombarded the city, and then captured 10 Indian ships and executed their crews. The Portuguese left Calicut bound for Cochin, farther south, where their arrival was peaceful. They loaded their six remaining ships with spices and sailed for home. Two ships were lost on the way and Cabral finally reached Portugal on June 23, 1501, with only four of the original 13 vessels.

Despite the many mishaps, King Manuel was pleased with the outcome of the expedition and for a time it seemed that Cabral might lead the next voyage to India, but the position went to Vasco da Gama. Cabral no longer held any position of importance at court and he retired to his estate, where he died. He was buried at the monastery at Santarém, in a tomb that was rediscovered in 1848.

FERDINAND MAGELLAN, FROM ATLANTIC TO PACIFIC

By the 16th century seafarers knew that the Earth was spherical, but it is one thing to know something because it makes sense, and something else to prove it. The only way to prove that the world is round is to travel all the way around it. The first expedition to achieve this, and supply the needed proof, was led by Ferdinand Magellan (ca. 1480–1521), although Magellan died before he could complete it.

Proving the shape of the Earth was not the purpose of the voyage. Magellan was seeking a route to the Spice Islands—the Moluccan Islands, now part of Indonesia. The Venetians and Genoese controlled the spice trade in the eastern Mediterranean and Muslim rulers held the lands to the east, but it should be possible to reach the Spice Islands by sailing west, thus avoiding both the Italians and the Muslims. It soon became evident, however, that whatever riches the Americas might contain, they had no spices.

Apart from the possibility of access to sources of valuable spices, Portugal was fighting to contain Muslim power in Africa and India, and in 1505 Magellan joined an expedition led by Francisco de Almeida (ca. 1450–1510) that fought a Muslim fleet off the Malabar Coast of India. On February 2–3, 1509, Magellan also fought in the Battle of Diu, in which the Muslim forces were defeated, and in June 1511 he took part in the battle that gave the Portuguese control of Malacca and unopposed access to Malaysia. They still had not reached the Moluccan Islands, however. Portuguese ships achieved this in 1512, but there is no proof that Magellan was present. He was back in Lisbon later that year, and fought at Azamor, in Morocco, where he sustained a wound that left him with a permanent limp. In November 1514, once more in Lisbon, he petitioned the king for a small increase in pay that would denote a promotion in rank. The king refused and sent him back to Morocco. Magellan tried again in 1516, but this time not only did the king refuse, he told Magellan he was free to offer his services elsewhere. Magellan had quarrelled with the commander of Portuguese forces in Morocco and had quit the army without permission, and Almeida had also made allegations against him. Clearly, Magellan had lost favor with the king. So he went to Spain, arriving in Seville on October 20, 1517. Also in 1517, Magellan married Beatriz Barbosa, the daughter of an important official in Seville.

Fernando Magellan (or Hernando de Magallanes) was his Spanish name. He was born Fernão de Magalhães, in either Sabrosa or Porto, Portugal. His parents, Rui de Magalhães and Alda de Mesquita, had been minor nobles, and after their death, the 10-year-old Fernão had become a page to Queen Leonor at the court in Lisbon.

In Seville, Magellan met the Portuguese cosmographer Rui Faleiro and they traveled together to the royal court at Valladolid, where they both renounced their Portuguese nationality and swore allegiance to Spain. They then offered their services to King Charles I (1500–58), who was also the Holy Roman Emperor Charles V. Together, Magellan and Faleiro offered to prove by sailing west that the Moluccas lay to the west of the demarcation line defined in the 1493 papal bull and, therefore, that Spain was entitled to claim them. The king approved their plan on March 22, 1518, and Magellan and Faleiro were appointed joint captains general of the expedition, with the promise that they and their heirs would govern any lands they discovered and that they would receive 5 percent of the profits from the enterprise. In the end, Faleiro fell ill and was unable to sail with Magellan.

Magellan believed he could find a strait through South America that linked the Atlantic and Pacific Oceans, and that this strait lay far to the south. If the Spanish could discover such a strait, they would have access to the Pacific without having to pass the Cape of Good Hope, which Portugal controlled. On August 10, 1519, Magellan sailed down the Guadalquivir River in command of a fleet of five ships: the *Trinidad, San Antonio, Concepción, Victoria,* and *Santiago.* After waiting for five weeks at Sanlúcar de Barrameda, they departed from Spain on September 20. They reached Tenerife on September 26, and on October 3 they set sail for Brazil. Though becalmed and beset with storms, the ships finally entered the Bay of Rio de Janeiro on December 13. They continued southward, reaching the mouth of the Río de la Plata, which they explored in the hope that it might be the strait, and on March 31, 1520, Magellan was at Puerto St. Julian, at 49.33° S, where they established a settlement.

At midnight on April 2 two of the captains mutinied against Magellan. The mutiny failed because most of the crew remained loyal. Magellan quelled the rebellion, executing the captain of the *Concepción* and marooning the captain of the *San Antonio* and a priest when the fleet left Puerto St. Julian on August 24.

The *Santiago* was sent ahead to reconnoitre the coast, but was wrecked in a storm. All of the crew survived and two returned to Puerto St. Julian with the news of what had happened. That is why Magellan decided to delay the departure. On October 21 the fleet reached Cabo Virgenes (Cape of the Virgins) at latitude 52.8° S, where they found deep saltwater extending far inland. On November 1, All Saints' Day, four of the ships sailed into what they found to be a channel 373 miles (600 km) long. Magellan called the channel the Estrecho de Todos los Santos (All Saints' Channel). It was later renamed the Strait of Magellan. Ordered to explore the strait, the *San Antonio* deserted and returned to Spain, taking most of the fleet's provisions with it. On November 28 the remaining three ships, *Trinidad, Concepción,* and *Victoria,* emerged from the western end of the strait and into the "Sea of the South." Magellan wept with joy. The sea they entered was so calm that they named it the *Mar Pacifico*—Pacific (Peaceful) Ocean.

The three ships sailed northward, parallel to the coast, until December 18, when Magellan ordered a new course, heading northwest. By the time they next saw land, on January 24, 1521, the crews were reduced to chewing leather and eating biscuits fouled by rats and sawdust, and they were tortured by thirst. They crossed the equator on February 13, at approximately 158° W, and made landfall at Guam on March 6. They had been without fresh food for 99 days.

Magellan had assured the king that he knew the way to the Moluccas, so it was probably in order to take on supplies that instead of making for them directly, on March 9 he sailed southwest to islands that were later named the Philippines, arriving on March 16. Magellan made an alliance between the ruler and Spain and converted several of the most important men to Christianity. A few weeks later, however, Magellan and his men became involved in a dispute with the ruler of Mactan Island and Magellan was killed in battle on April 27.

Ferdinand Magellan was the first European to reach the Philippines and the first navigator to sail through the strait that now bears his name. He was the first to see Tierra del Fuego (Land of Fire), which he named because of the many campfires that were visible as the ships sailed through the strait. He planned the expedition and supervised its execution, but he did not live to complete it.

JUAN SEBASTIÁN ELCANO, THE FIRST CIRCUMNAVIGATOR

With Magellan dead, Juan Sebastián Elcano (1486–1526), also called del Cano, assumed command of the three remaining ships, the *Concepción,* the *Trinidad,* and the *Victoria.* They had lost so many men in the battle at Mactan that there were now too few to crew all three ships, so del Cano decided to abandon the *Concepción* and burn it. Elcano had taken part in the Puerto St. Julian mutiny, but Magellan had pardoned him and after keeping him in chains for five months had made him captain of the *Concepción.* He was a very experienced commander.

They sailed to Palawan Island in the Philippines and departed from there on June 21, bound for Borneo and guided by pilots familiar with the shallow waters in that region. They spent a month in Brunei and finally reached the Moluccas on November 6. They loaded their ships with spices and on December 21 they set sail for Spain, but before long the *Trinidad* began taking in water. The crew tried but failed to repair the leak, so it was decided the *Victoria* should continue to Spain across the Indian Ocean and round the Cape of Good Hope while the *Trinidad* attempted to return across the Pacific. More than half the crew died from disease and the survivors were forced to return to the Moluccas, where they were captured and imprisoned by the Portuguese. Their ship was later lost in a storm.

The *Victoria* arrived in Spain on September 6, 1522. Of the 240 men who had sailed with Magellan only 17 made it back to Spain, together with four natives of Timor. All the survivors were very weak. The king rewarded Elcano with an annual pension and a coat of arms showing a globe with the motto *Primus circumdedisti me* (You were the first to encircle me).

Elcano was born in 1486 in the town of Getaria, in the Basque region of Spain and in his early years he was an adventurer. He fought with Spanish forces in Italy and in Algiers, and later settled in Seville and became a captain and owner of a merchant ship. He broke the law by surrendering his ship to Genoese merchants to whom he owed money and asked the king for a pardon. The king granted it, but on condition that Elcano sail as sailing master on the *Concepción.* After returning from the Magellan expedition, in 1526 Elcano was appointed chief navigator on another expedition of seven ships, led by García Jofre de Loaísa (ca. 1490–1536), that planned to follow Magellan's route and establish a Spanish colony in the Moluccas. Only one of the ships reached the Moluccas and most of the sailors died from malnutrition, including Elcano, who died on August 4, 1526.

SIR FRANCIS DRAKE AND THE DRAKE PASSAGE

To the south of Tierra del Fuego and the Strait of Magellan there is another sea channel, between Cape Horn and the South Shetland Islands. This channel marks the boundary between the relatively mild climate of southern South America and the much more severe conditions encountered in Antarctic waters. Called the Drake Passage, it is 600 miles (1,000 km) wide and links the Atlantic and Pacific Oceans. It bears the name of Sir Francis Drake (ca. 1540/43–96), but in fact the first explorers to sail through it were Flemish and led by the Dutch explorer Willem Corneliszoon Schouten (ca. 1567–1625), accompanied by Jakob Le Maire (ca. 1585–1616). An experienced sailor, Le Maire was also the son of the Amsterdam merchant sponsoring the expedition. It was Schouten who gave the southernmost tip of the continent its name of Kaap Hoorn—Cape Horn—named for Hoorn, the town of his birth.

The expedition had two aims. One was to discover the Terra Australis, the southern continent that many scholars believed must exist in order to balance the landmasses of the Northern Hemisphere. That part of the expedition failed. The second objective was to find an alternative route to the Spice Islands, because by that time the United East India Company (VOC) held a monopoly on all trade passing through the Strait of Magellan and imposed many restrictions on other vessels. Schouten and Le Maire left Amsterdam in 1615 with two ships, but because the smaller of them caught fire and had to be abandoned, only one vessel sailed through the Drake Passage. Schouten visited a number of islands before arriving at Batavia, Java (now Jakarta, Indonesia) in October 1616. The Dutch governor refused to believe Schouten had discovered a new passage between the two oceans. He charged him with breaching VOC regulations, confiscated their cargo and sent Schouten and Le Maire back to the Netherlands. Le Maire died on the voyage.

Drake may not have sailed through the passage bearing his name, but he did sail around the world, between 1577 and 1580. Drake would have had access to maps that were much better than those available to his predecessors. Most notably, the Flemish cartographer Gerard Mercator (1512–94) had brought out his world map in 1569. Mercator had found a novel solution to the age-old problem of depicting the surface of a sphere—the Earth—on a flat sheet.

The simplest way to depict the Earth accurately is to make a globe. The ancient Greeks may have made globes, but the earliest one known was made in 1492 and it still exists. It is held at the National Museum in Nuremberg, Germany. The illustration on page 138 shows the globe, with Africa and Europe clearly visible; it was made by Martin Behaim (born 1436 or 1459, died 1507). Its *gores*—the sections of maps that fit together on the spherical surface of a globe—are drawn on parchment and the globe is 20 inches (50.7 cm) in diameter. Behaim was born in Nuremberg, but lived and worked in Flanders and Portugal, where he died. He made his globe while visiting Nuremberg. The task took two years and he nicknamed it his *Erdapfel*—"earth apple," an old name for a potato.

Globes are all very well, but navigators need maps or charts that they can spread across a table, and they need to be able to plot a course that does not involve constant changes of heading. The problem is that although lines of latitude are parallel, lines of longitude

The world's oldest surviving globe, made in 1492 by Martin Behaim *(Lauros/Giraudon/Bridgeman Art Library)*

converge at the poles, which means that in order to travel in a straight line, a navigator sailing in any direction other than due east or west must cross each meridian of longitude at a different angle. That is the problem Mercator solved. He devised a way of depicting the surface features of the world as a map in which the meridians of longitude are parallel. Using a Mercator map, the navigator could set a course drawn as a straight line that crossed all the meridians at the same angle. The line of such a course is called a *rhumb line* or *loxodrome.* It is convenient for the navigator, but a rhumb line is not the shortest distance between two points on the surface of a sphere. When drawn on a sphere, a line crossing all meridians at the same angle describes a spiral curving toward the pole, so the ship is following a curved path.

Mercator never revealed his method—and he lived before the invention of the calculus for dealing with constantly changing values—but he found a way to increase the distance between meridians with increasing distance from the equator. One effect of this is that on a Mercator map centered on the equator, the distance between meridians at the poles, where they converge, becomes infinite, making it impossible to depict the poles because they are spread across the top and bottom of the map. A Mercator map can be centered on any latitude, although most center on the equator, and the map distorts the size and shape of lands away from those latitudes. That is because increasing the distance between meridians also necessitates increasing the distance between parallels of latitude by an equivalent amount. The illustration on page 139 shows a map compiled in 1589 by Mercator's youngest son Rumold (ca. 1541–1600) based on the world map drawn by his father in 1569.

Mercator was born on March 12, 1512, at Rupelmonde, Flanders, where his parents were visiting his brother Gisbert, who raised the boy after his parents died. Gisbert arranged for his education at 's-Hertogenbosch and the University of Leuven, and it was while he was at 's-Hertogenbosch that he translated his name from Gheert Cremer to Gerardus Mercator, its Latin equivalent (*cremer* and *mercator* are the Dutch and Latin words for merchant). Mercator became highly skilled as a maker of scientific instruments and mapmaker, finally settling in the German town of Duisberg, where he established a successful business and where he died on December 5, 1594.

Francis Drake was born in Tavistock, Devon, in about 1540 or 1543, and went to sea when he was 18, sailing in merchant ships that augmented their income by piracy. In 1572 he obtained from Queen Elizabeth (1533–1603) a letter of marque (see "Sea Rovers, Pirates, and Privateers" on pages 73–76) authorizing him to plunder at will in the Spanish colonies. On one occasion in the course of this highly profitable operation, he crossed the Isthmus of Panama and set eyes on the Pacific, then barred to all but Spanish ships. Drake prayed for an opportunity to sail on that ocean.

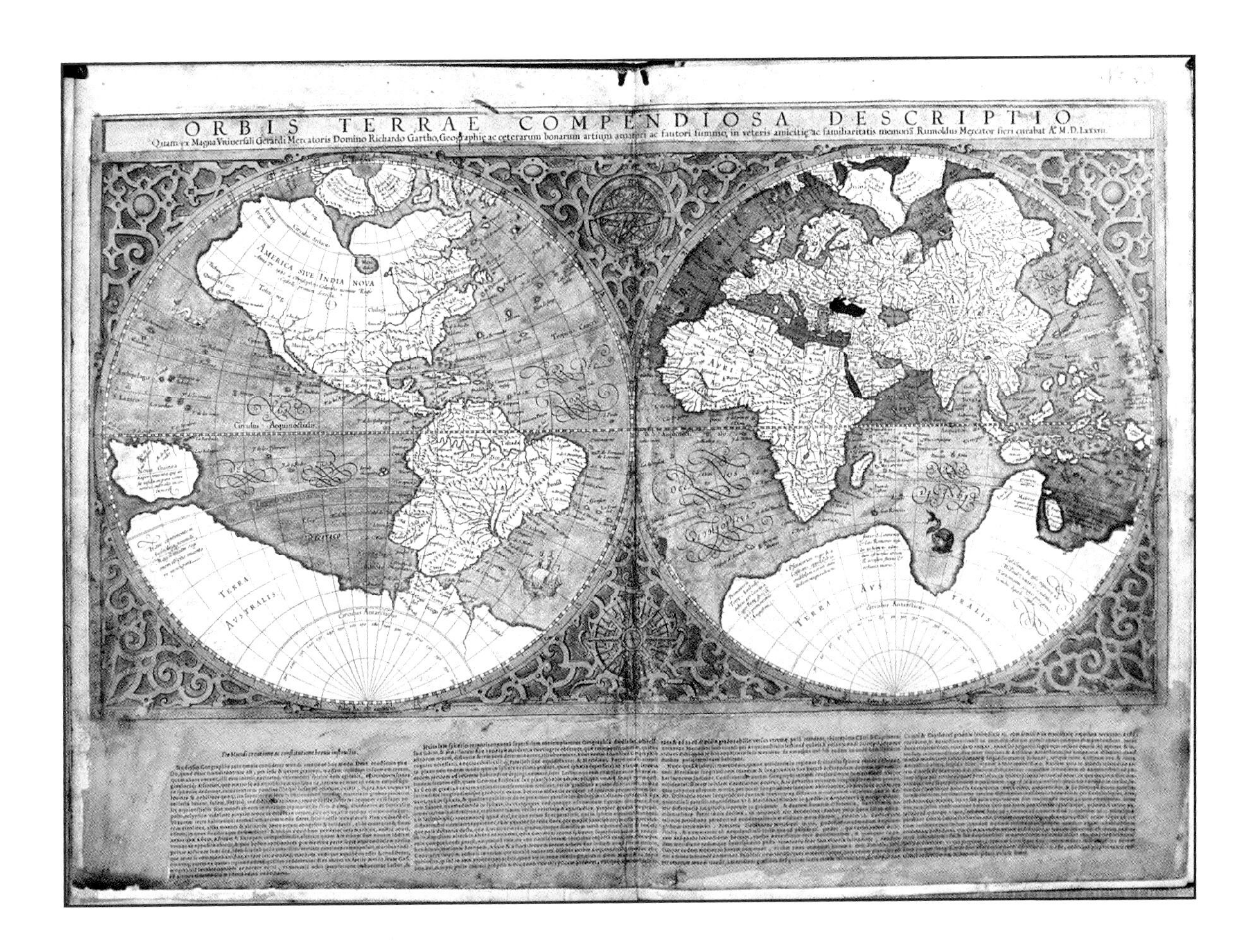

A map of the world compiled in 1589 by Rumold Mercator (ca. 1541–1600) and based on the world map drawn by his father, Gerard Mercator (1512–94), in 1569. This map is held at the Staatsbibliothek zu Berlin, Germany. *(Bildarchiv Preussischer Kulturbesitz/Art Resource, NY)*

His opportunity came when in 1577 he was chosen to lead an expedition through the Strait of Magellan. He had the queen's backing to make what profit he could and to cause as much damage as possible to the Spanish. Drake interpreted this as authorization for an extended voyage of piracy. He set sail from Plymouth on November 15, 1577, but encountered bad weather and was forced to return. He set sail again on December 13 on the *Pelican,* with five other ships and a total of 164 men. He later captured a Portuguese merchant ship, the *Santa Maria,* renamed it the *Mary,* and added it to his fleet. Many of his sailors perished crossing the Atlantic and after distributing their stores among the other ships Drake was forced to scuttle two of his ships, the *Christopher* and the *Swan,* because there were too few men to crew them. The *Mary* also had to be abandoned when its timbers were found to be rotten. The three surviving ships reached Puerto San Julian (see "Ferdinand Magellan, from Atlantic to Pacific" on pages 132–135) in the spring of 1578, where Drake saw the remains of the men Magellan had executed. Drake also faced a possible mutiny by disloyal officers and had one of them executed. His fleet entered the Strait of Magellan on August 21, emerging into the Pacific 16 days later and immediately sailing into ferocious storms. One of the three ships was destroyed and the other two were separated. Drake's second in command assumed the *Pelican* was lost and returned to England, leaving only the *Pelican* to continue the expedition. It is unlikely that Drake ever sailed through the Drake Passage.

Drake sailed north along the South American coast in the *Pelican,* which he now renamed the *Golden Hind,* raiding Spanish settlements that were largely undefended (hostile vessels had not been known to enter these waters). He sacked the port of Valparaíso for supplies and off the coast of Peru he captured two Spanish ships laden with gold, silver, and jewels. This capture revealed that the Spanish were active in Asia, which was strictly Portuguese territory under the Treaty of Tordesillas, and led to the Spanish invasion of Portugal in 1580.

The *Golden Hind* continued northward. Drake claimed to have reached the Aleutian Islands on his way to search for the Northwest Passage (see "Franklin, McClure, and the Discovery of the Northwest Passage" on pages 159–161), but the cold drove him back and he sailed south, finding good anchorage off land that he named New Albion and took possession of in the name of Queen Elizabeth. He kept the location of New Albion secret and historians have failed to identify it.

Drake headed west in July 1579 and sighted a group of islands, probably the Palau group, after 68 days. He called at the Philippines for freshwater and then sailed to the Moluccas, where he was received well and purchased spices. He called next at Java, sailed from there to the Cape of Good Hope, crossed the Atlantic, and arrived at Plymouth on September 26, 1580, laden with treasure and spices.

Drake's fortune was made. The queen knighted him in 1581 on board the *Golden Hind,* then lying at Deptford near London. He was made mayor of Plymouth and organized a water supply that served the city for the next 300 years. Drake continued to fight the Spanish and was admiral of the English fleet that defeated the Spanish Armada attempting to invade England in 1588. He died from dysentery on January 27, 1596, while anchored off Portobelo, Panama, and was buried at sea in a lead coffin, at his own request dressed in full armor.

ABEL JANSZOON TASMAN, WHO DISCOVERED TASMANIA AND NEW ZEALAND

In 1642 the governor-general of the Dutch East Indies was Anthonio van Diemen and when, on November 24 of that year, possibly the greatest of all Dutch explorers sighted a previously unknown land he called it Van Diemen's Land. The name was changed later and it is now called Tasmania after its European discoverer, Abel Tasman (1603–59). Tasman was also the first European to sight the Tonga Islands, the Fiji Islands, and New Zealand. He was also the first person to sail all the way round Australia.

Abel Janszoon Tasman was born in 1603, at Lutjegast, near Groningen in what is now the Netherlands. He went to sea at a young age and in 1631, when he married for the second time, he gave his occupation as sailor. He entered the service of the United East India Company (VOC) in 1632 or 1633 and in 1634 he made his first voyage on behalf of the company as captain of the *Mocha.* This was the first of several company ships he commanded on voyages to the East Indies (now Indonesia). He returned home in 1637, but sailed for the Pacific again in 1638, taking his wife with him.

In June 1639 van Diemen sent him on a voyage to the northwestern Pacific, in search of islands somewhere to the east of Japan that were reputed to be rich in precious metals. He never found them, but Tasman visited the Philippines, mapped a number of islands to their north, and visited Formosa (Taiwan). Tasman made several other voyages in the Pacific before heading south in 1642, in command of the *Zeehan,* in search of the southern continent. That was the voyage on which he discovered Tasmania. He then continued sailing eastward until, on December 13, his crew sighted land that he named Statenland, a name that was later changed to New Zealand. On January 21, 1643, the *Zeehan* sailed into the Tonga archipelago, where the crew obtained supplies. From there they sailed to the north of New Guinea and arrived back at Batavia (Jakarta) on June 15. The following year Tasman led an expedition of three ships that explored the northern coast of Australia, including the Gulf of Carpentaria, which Tasman named. His final voyage on behalf of the VOC began in 1648, when he led a fleet of eight ships in an attack on the Spanish. The attack failed, and the following year Tasman was dismissed. He was reinstated in 1650, but resigned in 1651. Tasman died in Batavia on October 22, 1659.

JAMES COOK AND SCIENTIFIC EXPLORATION

By the 18th century European governments and merchants needed more detailed and accurate charts of the oceans and maps of coastlines. Obtaining these involved long, costly, and frequently dangerous voyages that had to be paid for from public funds. The task was entrusted to navies and naval surveying missions became increasingly scientific. James Cook (1728–79) was probably the first of the truly scientific explorers, navigators, and chart makers.

James Cook was born on October 27, 1728, in the village of Marton-in-Cleveland, in North Yorkshire, England, one of the five children of James Cook, a Scottish farm laborer, and his wife Grace, a local woman. Soon after James was born his father obtained a job as foreman on a farm at Great Ayton, on the edge of the North York Moors, owned by Thomas Scottowe, and that is where James spent his childhood. Scottowe paid for him to attend the village school. In 1745, Cook was apprenticed to William Sanderson, a grocer and haberdasher, but he was strongly attracted to the sea and Sanderson introduced him to John and Henry Walker, ship owners and coal shippers in the nearby port of Whitby. In July 1746 James's apprenticeship was transferred to them. In summer he would sail on colliers between ports along the eastern coast of England and in winter he remained ashore, studying navigation and mathematics. His apprenticeship completed, James worked on ships trading with the Baltic Sea ports, returning to the Walkers in 1752 with the rank of mate. In 1755 the Walkers offered him his own command, but Britain was preparing for war with France—it would become the Seven Years' War, also called the French and Indian War—and James enlisted in the Royal Navy. On June 17 he began his naval service as an able seaman and within a month he had been promoted to master's mate. By 1757 he had passed the examination qualifying him as a ship's *master* and was appointed master of the 64-gun HMS *Pembroke.* In the days of sail a ship's master was in charge of the ship's navigation and steering. He ordered the amount of sail to be carried, but he was not in overall command of the vessel.

In February 1758 the *Pembroke* took part in the siege of Louisburg, Nova Scotia, and the assault on Quebec, where James Cook played an important part in the charting of the St. Lawrence. After the battle he was transferred to HMS *Northumberland* as master and while on the *Northumberland* Cook improved his knowledge of mathematics and astronomy and also learned surveying. In 1763, after a short spell ashore, he was appointed to the Newfoundland survey as master of the schooner *Grenville.* He spent five summers surveying the coast and the winters ashore working on his charts.

In 1766 Cook observed a solar eclipse from the Burgeo Islands off the Newfoundland coast. His observations and his use of them to calculate his longitude were published in the *Philosophical Transactions of the Royal Society* for 1767, with a comment complimenting Cook on his mathematics. Cook had attracted the attention of the Royal Society and also of the Admiralty, and in 1767 he was commissioned as a lieutenant and given command of His Majesty's Bark *Endeavour,* a collier bought by the navy to carry observers to Tahiti, where they were to witness a transit of Venus across the Sun on behalf of the Royal Society. James Cook and the astronomer Charles Green (1735–71) were to be the two observers. This was the first of what were to be Cook's three voyages of discovery in the Southern Hemisphere.

The *Endeavour* sailed from Plymouth on August 25, 1768 and reached Tahiti on April 13, 1769. The observations of the transit apparently went well, but all observations of Venus were uncertain and the calculations based on them varied so widely that the exercise was useless. Nevertheless, Cook established friendly relations with the Tahitians and two of his passengers, the Swedish botanist Daniel Carlsson Solander (1733–82) and the English botanist Sir Joseph Banks (1743–1820), were able to collect many plant samples. The *Endeavour* left Tahiti on July 13, visiting and naming the Society Islands (the name refers to the way the islands are grouped together).

Cook was also charged with finding the southern continent. When he failed to do so he headed for New Zealand, arriving on October 7, then sailed all the way around the New Zealand coast in a figure eight, mapping the entire coastline and making friends with the New Zealanders. Next he headed for Australia, arriving on April 19, 1770. Cook and members of his crew went ashore on April 29 and Cook named that stretch of coast Botany Bay because of the many unique specimens the two botanists gathered. The *Endeavour* then headed northward, but the ship ran aground on the Great Barrier Reef and spent several weeks undergoing repair. They passed through the Torres Strait to Batavia, where malaria and dysentery broke out among the crew, with many fatalities. Cook brought the ship home, arriving back at Plymouth on July 12, 1771. He was immediately promoted to the rank of commander.

Determined to establish once and for all whether a southern continent existed, Cook proposed a second voyage and the Royal Society commissioned him to undertake it with two ships, HMS *Resolution* and HMS *Adventure.* They departed from England on July 13, 1772. Cook crossed the Antarctic Circle on January 17, 1773, reaching 71.17° S. He mapped South Georgia island and explored the South Sandwich Islands, but failed to find the continent, although he remained convinced that it must exist somewhere to the south of the sea ice. On October 30, 1773, the two ships parted company when they were separated in fog, and the *Adventure* returned to Britain. Between February and October 1774 Cook sailed across the Pacific, visiting many smaller islands. During this voyage Cook used a copy of the chronometer made by John Harrison (see "John Harrison and his Time-keeper" on pages 102–105). His logbooks were full of praise for this instrument and the charts he produced were so accurate that some remained in use for almost two centuries.

The *Resolution* arrived back in England on July 30, 1775. Cook was promoted to the rank of *post captain*—a naval rank, as opposed to a courtesy title used of anyone commanding a ship—and given an honorary retirement, with an administrative position at the Navy's Greenwich Hospital. He was elected a fellow of the Royal Society, which awarded him its Copley Medal for a paper he wrote on the methods he had used to combat scurvy. Cook could have lived out the rest of his life in comfort, but soon he was planning a third voyage, this time to the north, to search for the Northwest Passage.

Cook set sail on July 12, 1776, once more in command of HMS *Resolution,* and headed south. HMS *Discovery* joined him at the Cape of Good Hope and the two ships continued to Van Diemen's Land (Tasmania), New Zealand, Tonga, and the Society Islands before turning northward on December 8, 1777. On January 18, 1778, the expedition discovered the Hawaiian Islands, where they went ashore at Waimea Harbor, Kauai. Cook named the island group the Sandwich Islands, in honor of the earl of Sandwich (1718–92), who was acting as First Lord of the Admiralty. The name fell into disuse during the 19th century.

After leaving Hawaii the ships headed northeast until they reached California, then explored and charted the western coast of North America as they sailed northward. They sailed around the Aleutian Islands and through the Bering Strait, reaching latitude 70.73° N before the pack ice forced them to turn back. Cook determined to spend the winter at Hawaii. After surveying part of the coast, the *Resolution* and *Discovery* anchored in Kealakekua Bay, Big Island, on January 17, 1779. They left Hawaii on February 4, but had to return for repairs. On February 14 some Hawaiians stole one of the *Resolution*'s small boats. This was a fairly common occurrence, dealt with by taking hostages and holding them until the stolen property was returned. Cook led a party ashore intending to capture a hostage, but the Hawaiians resisted and the men were forced to retreat to the beach. As Cook was helping launch the boats the pursuing Hawaiians struck him on the back of the head, then stabbed him to death and dragged his body away. Four of the marines in the party were also killed and two were wounded.

James Cook had married Elizabeth Batts (1742–1835) on December 21, 1762. They had six children, three of whom died in infancy. Two of his three surviving sons joined the navy, but all three were dead by 1794.

VITUS BERING, WHO DISCOVERED ALASKA AND THE BERING STRAIT

Vitus Jonassen Bering was a Danish navigator, born on August 12, 1681, in the city of Horsens in East Jutland, Denmark. In 1703 he joined the Russian navy as a sublieutenant during the reign of Peter the Great (1672–1725). He was promoted to lieutenant in 1704 and fought in the Russo–Turkish (Great Northern) War of 1710–11. Czar Peter was interested in establishing Russian colonies in North America and in finding a northeast passage—that is, a sea route to China around northern Siberia—and in January 1725 he placed Bering in charge of the first expedition to determine the extent of the mainland of Siberia and to discover whether Siberia and North America were linked. Bering led the party more than 6,000 miles (9,650 km) along the north coast of Russia, reaching the Pacific coast on September 30, 1726; there, the team built ships that carried them to the Kamchatka Peninsula. Bering then explored the coast, sailing northward as far as 67.3° N and passing through the strait that now bears his name. Bering did not sight the North American coast and was convinced that the two continents were not joined, but when he reported this to the Russian admiralty the officials found his argument unconvincing.

Bering then proposed a much larger expedition to settle the matter. The plan was agreed in 1732 and he was placed in command of the Great Nordic Expedition. Preparations were completed in February 1733. The party left St. Petersburg, crossed Siberia and finally reached Okhotsk on the Pacific coast in 1735. Bering commissioned local shipbuilders to construct two vessels, the *Sviatoi Piotr (St. Peter)* and the *Sviatoi Pavel (St. Paul).* The expedition sailed to Kamchatka, where in 1740 they established the settlement—now a city—of Petropavlovsk. In 1741 the two ships sailed together for North America with Bering in command of the *Sviatoi Piotr* and his second-in-command Alexei Chirikov (1703–48) in command of the *Sviatoi Pavel.* The ships became separated during a storm, but in July 1741 Chirikov and his crew sighted the southeastern coast of Alaska and anchored off Cape Addington at 58.47° N, and a few days later Bering's party sighted Mount St. Elias. On their return journey both ships explored several of the Aleutian Islands.

Chirikov and the crew of *Sv. Pavel* returned safely to Siberia, but Bering fell ill and had to relinquish command of his ship, which was wrecked on what is now called Bering Island, the largest of the Commander Islands (Komandorskiye Ostrova), where Bering died from scurvy on December 19, 1741, along with most of his crew. Helped by the other survivors, the only surviving carpenter, S. Starodubtsev, managed to build a smaller boat from the wreckage, and in this boat 45 of the original crew of 77 returned to Russia. In addition to Bering Island and the Bering Strait, the Bering Sea, Bering Glacier, and Bering Land Bridge bear the explorer's name.

THE VOYAGES OF CHRISTOPHER COLUMBUS

Shortly before dawn on August 3, 1492, Christopher Columbus (1451–1506) set sail from Palos, a port on the Gulf of Cádiz, Spain, at the commencement of the most famous sea voyage in history. The expedition consisted of three caravels. Columbus, the leader, commanded the *Santa María,* Martín Alonso Pinzón commanded the *Pinta,* and Martín's brother Vicente Yáñez Pinzón commanded the *Niña.* They reached the Canary Islands on August 12 and departed from there on September 6, heading westward into the open Atlantic.

Christopher Columbus is the name by which the explorer is known in the English-speaking world. One of the four sons of a weaver, he was born in Genoa, Italy, possibly into a Spanish-Jewish family, as Cristoforo Colombo. He spent time in Portugal, where he used the Portuguese version of his name, Christovão Colom; when he entered the service of Ferdinand and Isabella, the king and queen of Spain, his name became Cristóbal Colón. Columbus had first gone to sea at the age of 14 and by the time of his Atlantic voyage he was a highly experienced seafarer and a skilled navigator. He may have sailed on a privateering ship and in 1476 he was sailing on one of five Genoese ships bound for England that were attacked by privateers and Columbus' ship was sunk. He swam six miles (10 km) to the Portuguese shore clinging to wreckage.

His aim was to reach the spices, gems, and other riches of Asia by sailing westward. A deeply pious man, Columbus found justification for his plans in various scriptural passages that he interpreted as predictions of success. His extensive reading of the accounts of travelers, as well as of the Bible, led him to conclude that the surface of the spherical Earth is covered by six parts dry land and one part ocean, and that the distance between Spain (the edge of the West) and India (the edge of the East) was very long by land but very short by sea. Columbus reckoned that traveling eastward by land across Europe and Asia, the distance between Spain and India was 282° of longitude. There are 360° of longitude in all, so Columbus surmised that the distance between Spain and India traveling westward by sea must be (360° - 282°) = 78°. That being so, Columbus calculated the distance to be 2,760 miles (4,440 km).

Columbus had a map to help him prepare. The original version had been drawn by Ptolemy (Claudius Ptolemaeus, ca. 90–ca. 168 C.E.), an astronomer and geographer, probably Egyptian, who lived in Alexandria. The map had appeared in Ptolemy's book *Geographia,* but Italian cartographers had subsequently greatly modified it. The map suggested the possibility of reaching India by sailing westward. The original is lost, but the illustration on page 149 shows a 16th-century copy. Columbus also had a chart to help him navigate prepared by Paolo Toscanelli (1397–1482), a Florentine physician and mapmaker. Toscanelli based his chart on Ptolemy's map, embellished it with travelers' tales and legends, and showed the Atlantic Ocean with Europe in the east and Asia in the west.

As the days dragged on and the three ships continued westward, Columbus realized they must have already covered about 5,175 miles (8,327 km) rather than the 2,760 miles (4,440 km) he had anticipated, which suggested that the Earth must be larger than it appeared on his chart. Nevertheless, when his increasingly hungry, scared, and mutinous crew finally espied land, two hours after midnight on October 12, Columbus had not the slightest doubt where they were. He named the first island they reached San Salvador, claimed it for Spain, and was convinced it was one of the outlying islands close to Cipango (Japan). The island he named San Salvador was Guanahani, in the Bahamas, and the people he met were definitely not Japanese.

Columbus was wrong on every count, but this was not his fault. It was the map. Ptolemy had calculated the distances on his map by using a few astronomical measurements, but he had also relied on reports he obtained from sailors about the time it took to sail between particular points. This was a type of dead reckoning, and much too approximate to be useful in drawing an accurate map. Ptolemy related these distances to the size of the Earth that had been calculated by the Greek philosopher Poseidonius (ca. 135–51 B.C.E.). Poseidonius had calculated the Earth's circumference by observing

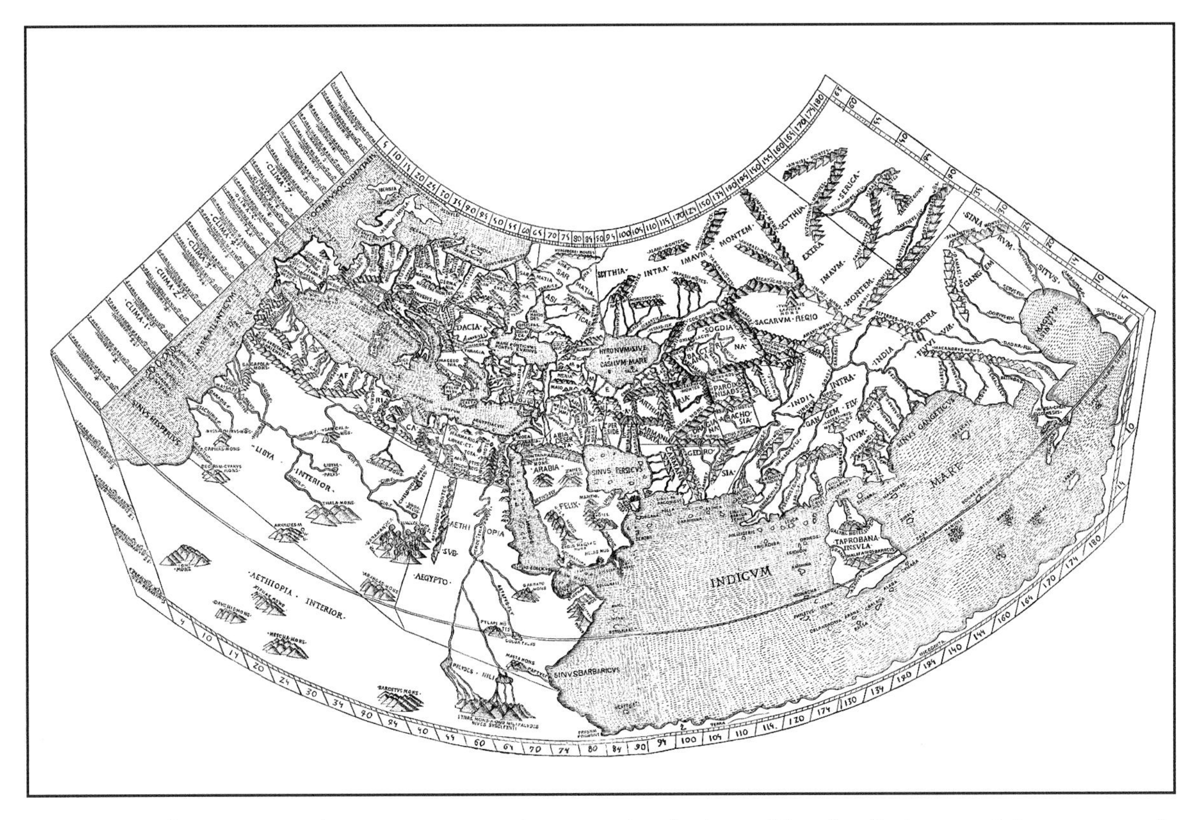

In the second century C.E., the astronomer and geographer Ptolemy (Claudius Ptolemaeus) drew a map of the known world extending from Britain to Asia. This is a hand-colored woodcut copy of that map, made in the 16th century. *(North Wind Picture Archives)*

the star Canopus. He noted that at Rhodes he could see Canopus on the horizon but never rising above it, while at Alexandria Canopus rose 7.5 degrees above the horizon. He believed the distance between Rhodes and Alexandria was 5,000 stadia, that the difference in the *declination*—the angle between the star and the horizon—of Canopus between these two places was 1/48th part of a circle (360 ÷ 7.5 = 48), so he multiplied 5,000 by 48 to reach a circumference of 240,000 stadia. No one is quite certain of the length of a stadion in modern units, but scholars believe it was between 500 feet (152 m) and 607 feet (185 m). That means Poseidonius's calculated circumference, of between 22,727 miles (36,568 km) and 27,590 miles (44,394 km), was very close to the true value of 23,674 miles (38,092 km). But then Poseidonius revised his estimate of the distance between Rhodes and

Alexandria from 5,000 to 3,750 stadia, thereby reducing the calculated circumference of the Earth to between 180,000 stadia—17,045 miles (27,426 km)—and 20,693 miles (33,295 km), which is between 6,629 miles (10,666 km) and 2,981 miles (4,796 km) short of the true value. Ptolemy used Poseidonius's revised value, and that is why Columbus and his crew ran out of supplies on a voyage that turned out to be much longer than his map indicated.

Ptolemy did not have to use the Poseidonius value. Eratosthenes (ca. 276–ca. 196 B.C.E.), the Greek philosopher who was one of the world's first geographers, had also calculated the Earth's circumference. Eratosthenes knew that near Syene (modern Aswān), Egypt, there was a deep well where on Midsummer Day the Sun at noon shone directly onto the water. Syene was very close to the tropic of Cancer. He then measured the declination of the Sun at noon on Midsummer Day at Alexandria by measuring the length of the shadow cast by an obelisk, the height of which he knew, and used trigonometry to calculate that the Sun was 7.2° from being directly overhead. Eratosthenes also knew the distance between Alexandria and Syene. Egyptian pacers (men who measured distances by pacing them out) and camel drivers had measured it as 5,300 stadia. Syene is not directly south of Alexandria, so Eratosthenes corrected the distance to 5,000 stadia. He calculated that 7.2° is 1/50th of a circle, so 5,000 × 50 = 250,000 stadia, which he rounded to 252,000 stadia so it would be divisible by 60. His value of between 23,863 miles (38,397 km) and 28,970 miles (46,613 km) for the Earth's circumference was remarkably close.

Columbus returned to Spain, arriving on March 15, 1493, to an ecstatic reception. He displayed the islanders he had brought with him and he dined at the same table as the king and queen. He was also rewarded financially and given the title of admiral of the ocean seas. His second voyage began on September 25, 1493, this time with 17 ships and more than 1,000 passengers who planned to settle in the new lands. His ships carried livestock, seeds, and building materials. They claimed Guadeloupe and Puerto Rico for Spain and reached Hispaniola (the island now containing Haiti and the Dominican Republic) in late November, only to find that the settlers left on the island from the first expedition had been killed, presumably by local people. His passengers set to work building a new settlement. The fleet went on to explore more of Cuba and Jamaica, returning to Spain in 1496.

On May 30, 1498, Columbus departed from Seville on a third voyage, with six ships, sending some of the ships to assist the settlers in Hispaniola while he led the others farther south than he had traveled previously. His party discovered and named Trinidad, then sailed to the South American coast where they found the mouth of the Orinoco River. On his return to Hispaniola, however, Columbus found that a quarrel had broken out among the settlers. He resolved it by granting plots of land, together with their indigenous inhabitants as slaves, to those settlers who had opposed the regime he had established. This took two years and it was November 1500 before he arrived back in Cádiz. Various accusations had been leveled against Columbus and his two brothers on behalf of the disgruntled settlers in Hispaniola. These were dropped only after the intervention of the king and queen, but not before Columbus had been replaced as governor of the Spanish colonies by a person sympathetic to the rebels.

The two monarchs directed Columbus to undertake a fourth expedition, this time explicitly in search of gold, silver, gems, spices, and other valuables. He sailed from Cádiz on May 9, 1502, accompanied by his brother Barthlomew and his 14-year-old son Fernando. Now aged 50, Columbus was in poor health and his eyesight was failing. He was not well enough to command a fleet of four ships and 150 sailors, but relied on seamen he knew well and who were loyal to him. The fleet took on supplies at Gran Canaria, in the Canary Islands, called at Martinique, and stopped at Santo Domingo, Hispaniola, for repairs and to shelter from a hurricane before proceeding to Central America. One ship was lost at sea, weakened by shipworms that had eaten into its timbers, and another sank close to the coast of Panama. The surviving 130 men crowded onto the remaining two ships. These were also badly damaged by shipworms and barely seaworthy. The captains realized they could not reach Hispaniola, so they headed for Jamaica where they beached the ships and abandoned them. They were marooned in Jamaica for more than a year, and rescued only after two of the party managed to reach Hispaniola, more than 100 miles (160 km) distant, in two canoes paddled by islanders. Columbus finally reached Santo Domingo and sailed for Spain, arriving on November 7, 1504.

Columbus was now very weak. He retired to a monastery near Seville where he spent several months recovering from the stresses and strains of his final voyage, then began following the king, writing letters and seeking an audience in the hope of regaining titles and claims to riches that had been stripped from him during the dispute with the settlers. In 1505 Ferdinand granted him 2 percent of the riches of the Indies. Now a wealthy man, Columbus remained in the capital, Valladolid, where he died on May 20, 1506.

The Arctic

Until the day he died Christopher Columbus remained convinced that the lands he had discovered on the far side of the Atlantic Ocean were in Asia. It was soon apparent, however, that he was mistaken. Columbus had visited islands and the mainland of a New World consisting of two vast and linked continents.

The discovery of the Americas did not mean that the idea of sailing westward to reach Asia was wrong, but it did mean that the Americas presented a very large obstacle in the way of realizing that dream. A European adventurer hoping to reach the Orient by heading westward would have to sail around the New World. Within a few decades of Columbus's first voyage, other explorers had rounded Cape Horn, and in 1520 Ferdinand Magellan had found the strait bearing his name (see "Ferdinand Magellan, from Atlantic to Pacific" on pages 132–135). They had proved the possibility of sailing from the Atlantic to the Pacific, but they had also shown that this involved a very long journey. Might it not be feasible to reach the Pacific by a much shorter northern route?

This chapter tells of the explorers who headed north. Most went in search of the Northwest Passage–the sea route linking the two oceans. Others aimed to reach the North Pole.

SIR MARTIN FROBISHER AND THE FIRST VOYAGES TO THE FAR NORTH

For anyone dreaming of the life of an explorer, the first hurdle to cross is not geographic, but economic. Exploration is an expensive business. Martin Frobisher (1535 or 1539–94) was in his early twenties and already an experienced seaman when he first declared his intention of seeking the Northwest Passage to the riches of Asia, but it took him 15 years to find backing for an expedition.

Frobisher was born in Yorkshire, England, into a large family; at a young age, he was sent to a school in London, where he lived under the care of a relative, Sir John York. In 1544 York sent the young man to sea on a ship that was part of a merchant fleet sailing to West Africa. He spent the next 20 years or so sailing mainly in the Mediterranean.

In 1576 Frobisher finally won backing for his voyage of exploration from the Muscovy Company, which had also financed expeditions to seek the Northeast Passage (see "The Northeast Passage" on pages 157–158). With three small ships under his command, and the good wishes of Queen Elizabeth to encourage him, Frobisher sailed from Greenwich on June 7, 1576, heading north through the North Sea to the Shetland Isles, to the north of mainland Scotland, and from there heading northwest. Two of his ships were lost in bad weather, but on July 28 the crew of the *Gabriel* came within sight of the coast of Labrador. They reached what became known as Frobisher Bay a few days later; finding it impossible to sail farther north, Frobisher determined to head west. On August 18 he reached Baffin Island, where five of his men were captured by Inuit people and never seen again, despite Frobisher's efforts to recover them. He headed for home and arrived in London on October 9.

Frobisher brought home with him a quantity of a black stone that he believed (or said he believed) contained gold. In fact it was fool's gold—pyrites (iron sulfide, FeS_2)—but with the help of the Muscovy Company, Frobisher was able to use it to win backing for a second voyage. This time the queen sold a naval ship to the expedition and provided a substantial amount of the expedition's finance. Frobisher departed from London on May 27, 1577, with three ships and a company of 150 men that included soldiers, miners, and persons who knew how to refine metal ores. Again they sailed across the north of Scotland; reaching Frobisher Bay on July 17, Frobisher claimed all the islands in the Queen's name. The expedition spent its time collecting what they thought was valuable metal ore before leaving for England on August 23. All three ships returned safely and the queen congratulated Frobisher personally. The supposed ore was sent for assaying.

There was still no gold, but the queen and Frobisher's other backers remained convinced of the potential riches to be won, and a third and much bigger expedition was organized. This time, when she received him to bid him farewell, Elizabeth placed a gold chain around his neck. The expedition sailed from Plymouth, departing on June 3, 1578, with 15 ships and enough supplies to establish a colony with 100 inhabitants in the new territory. The fleet reached southern Greenland on June 20 and on July 2 they arrived at Frobisher Bay, where they landed with some difficulty after bad weather had destroyed one of the ships. Frobisher tried to set up a colony, but quarrels among the colonists made the task impossible. After collecting a large quantity of ore the party headed for home, arriving early in October. The ore was taken to a smelting plant constructed especially to receive it, but assayers found it was not worth smelting. Eventually it was used in road making and the final acceptance of the fact that there was no gold to be found in Baffin Island ended Frobisher's voyages of exploration.

Frobisher had not fallen from favor, however. In 1580 he was made captain of a naval ship and in 1585 he was promoted to vice-admiral in command of the *Primrose,* accompanying Drake on his voyage to the West Indies (see "Sir Francis Drake and the Drake Passage" on pages 136–141). He commanded the *Triumph,* one of the ships that dispersed the Armada in 1588, and was knighted for his service in that battle. In November 1594 Frobisher suffered a gunshot wound in battle and he died in Plymouth on November 15.

HENRY HUDSON AND HIS BAY

The storms that Frobisher's party encountered when they reached Frobisher Bay on their final expedition drove the fleet about 60 miles (96 km) along an unexplored strait that later came to be named the Hudson Strait. Its name refers to the English explorer Henry Hudson (ca. 1565–1611), as do those of the Hudson River and Hudson Bay. Very little is known about Hudson's early life; even the year of his birth is uncertain. He may have been born in Hertfordshire, to the north of London, and several men called Hudson, who may have been relatives, were associated with the Muscovy Company, which sponsored Hudson's first explorations. What is known is that in 1585 a voyage that sailed in search of the Northwest Passage had been planned in the Limehouse, London, home of a Thomas Hudson, and Henry Hudson may have been present. Certainly Henry acquired a considerable knowledge of the Arctic before he sailed there.

In the spring of 1607 the Muscovy Company appointed Hudson to lead an expedition in search of the Northeast Passage (see "The Northeast Passage" on pages 157–158). Accompanied by his son John and 10 companions, Hudson sailed on May 1. When they reached the edge of the sea ice they followed it eastward until they arrived at the Svalbard Archipelago. Then they continued to investigate an area first explored by the Dutch navigator Willem Barentsz, also known as William Barents (ca. 1550–97).

In 1608 Hudson embarked on a second expedition on behalf of the Muscovy Company, this time to seek the Northeast Passage to the east of the Barents Sea. He reached the area between Svalbard and Novya Zemlya, but found impenetrable ice blocking his passage and was compelled to return to England. Soon after his return Hudson traveled to Amsterdam, where the Dutch East India Company (VOC) hired) him to undertake a third search for the Northeast Passage. While he was in Amsterdam, Hudson learned of two possible routes to Asia along channels to the north of North America, one at about latitude 40° N and the other at about 62° N. His commission was to search for a Northeast Passage, however, and he promised the VOC that he would return at once if he failed to find it.

On April 6, 1609, Hudson sailed from Amsterdam on the *Halve Maen (Half Moon).* Again he headed north of Russia, but he encountered storms and headwinds that forced him to turn back. Instead of returning to Holland, however, Hudson proposed to his crew that they should seek the Northwest Passage at latitude 40° N. They crossed the Atlantic, then sailed parallel to the North American coast until they came to the mouth of a major river (now known as the Hudson River) that had been discovered in 1524 by the Florentine navigator Giovanni da Verrazano (1485–1528). They sailed upstream for about 150 miles (241 km), to about where the city of Albany is now, before deciding that they were on a river, not a strait leading to the Pacific.

On his way back to Holland, Hudson called at Dartmouth, Devon, and the English government forbade him from undertaking any further voyages of exploration on behalf of other nations. His logbooks and other papers were sent to the VOC in Holland. Hudson now received funding from two English companies, the Virginia Company and the East India Company, as well as the Muscovy Company, and set out in search of the other possible strait, at about 62° N, which had been described in 1602 by another explorer, Captain George Weymouth. In his logbook Weymouth had described what he called a furious overfall of water that rushed out of a certain inlet at every ebb tide, suggesting that a large body of water—possibly the Pacific Ocean—lay beyond the inlet.

Hudson sailed from London on April 17, 1610, on the *Discovery.* He called at Iceland and then made his way to the "furious overfall"—Hudson Strait—and sailed through it to the open water beyond. He had entered what was later known as Hudson Bay, but instead of continuing in a westerly direction Hudson followed the coast southward to James Bay. He found no outlet to the Pacific and stayed too long in the bay. Winter overtook the expedition. Trapped, they moved ashore.

Quarrels then began. Hudson gave a gray gown to Henry Green, a member of the crew, then fell out with him, claimed the gown back, and gave it to another man. This caused resentment. Other crew members suspected Hudson of hiding food to give to his favorites, and Hudson had the crew's sea chests searched for hidden food. Finally, Green and the mate, Robert Juet, planned a mutiny. They sailed for home in June 1611, when the ice cleared, although Hudson had wanted to continue exploring. On June 22 the crew seized Hudson, his son, and seven men loyal to him and set them adrift in a small open boat with a few basic supplies. They were never heard of again. Both Green and Juet were later killed in a fight with Inuits. The *Discovery* returned to England and although eight of the 13 surviving mutineers were arrested, all were released.

THE NORTHEAST PASSAGE

Prior to the opening of the Suez Canal in 1869, ships sailing from ports in northern Europe had to travel round either the Cape of Good Hope or Cape Horn in order to trade in Asia. If a route from the Atlantic to the Pacific could be discovered that ran parallel to the north coast of Eurasia, it would obviously be much shorter, quicker, and, therefore, much cheaper. This would be a northeast passage to Asia and it would be of greatest benefit to Russia, which had no Atlantic ports. The Russian scholar and diplomat Dmitry Gerasimov (ca. 1465–ca. 1535) may have been the first person to alert others about the possibility of the existence of such a passage. While serving as ambassador in Rome, Gerasimov described the geography of Russia and the countries of northern Europe to the Italian historian Paolo Giovo (1483–1552). Giovo repeated the information in a book, and in 1525 the Genoese cartographer Battista Agnese (ca. 1500–64) drew a map based on it. Agnese's map showed that, apart from islands and peninsulas, there was no land to the north of the Eurasian north coast. It might be possible to navigate through the ice that covered the sea for much of the year.

Over the course of the following centuries several explorers sought the Northeast Passage and Vitus Bering sailed through it (see "Vitus Bering, who Discovered Alaska and the Bering Strait" on pages 146–147) in 1728. The first navigator to sail through the Northeast Passage successfully was the Finnish-born, naturalized Swedish explorer Adolf Erik Nordenskiöld (also spelled Nordenskjöld, 1832–1901). Nordenskiöld twice sailed across the Kara Sea, to the east of the large island of Novaya Zemlya, as far as the mouth of the Yenisei River, in 1875–76 and 1876–77.

On June 22, 1878, Nordenskiöld sailed from Karlskrona on the Baltic coast in southern Sweden on the steamship *Vega*. He reached the mouth of the Yenisei on August 6 and the mouth of the Lena, in the Laptev Sea, on August 27. The *Vega* became locked in ice on September 27, only about 100 miles (160 km) from the Bering Strait, and was unable to proceed until the following spring. Nordenskiöld sailed through the Bering Strait and into the Pacific in July 1879. He visited Alaska before returning to Sweden by way of Japan, the Suez Canal, and the Mediterranean. On his return to Stockholm in 1880, Nordenskiöld was made a baron, and in 1893 he became a member of the Swedish Academy.

The Northeast Passage has been used commercially since the early 1930s and it has remained open in summer since the late 1960s. Routine aerial reconnaissance has warned of its imminent closure by ice and icebreakers have then been deployed to maintain a clear route. Retreating sea ice left the passage open in the summers of 2005 and 2008. In 2009 the ice returned.

FRANKLIN, MCCLURE, AND THE DISCOVERY OF THE NORTHWEST PASSAGE

For the United States and Canada, as well as for the nations on the western edge of Europe, it is the Northwest Passage that offers the shortest sea route from the Atlantic to the Pacific. Explorers and navigators began searching for this route from about the year 1500, as soon as it became evident that the land Columbus had discovered was a vast continent lying between Europe and Asia—and not Asia itself, as Columbus had supposed. Navigating the Northwest Passage proved much more difficult and dangerous than sailing the Northeast Passage, however.

The Northwest Passage follows a route lying about 500 miles (800 km) inside the Arctic Circle. The north of Canada consists of many islands that extend from Baffin Island in the east to Banks Island in the west. A navigator has no alternative but to follow the deep-water channels through this maze of islands, sailing northwest through Baffin Bay that separates Baffin Island from Greenland, then passing to the north of Baffin Island, and finally emerging into the Beaufort Sea, to the north of Alaska. Permanent sea ice makes it impossible to miss the islands altogether by sailing to the north of them. Baffin Bay is open water, but there are many drifting icebergs that calve from the outlet glaciers flowing from the Greenland ice cap. Once past the islands, the Beaufort Sea contains ice that prevailing winds push in the direction of the Bering Strait.

For centuries the Northwest Passage existed as little more than a dream. No mariner had discovered it and no one could prove that it existed. The sailor who first proved its reality was Sir John Franklin (1786–1847), a rear admiral in the British navy. Franklin began by exploring the route in stages. Between 1819 and 1822 he led a party that sailed by canoe up the Coppermine River to the Arctic Ocean opposite Victoria Island, then eastward along approximately 210 miles (338 km) of coast to the western side of Hudson Bay. On a second expedition, from 1825 to 1827, Franklin and his team began at the mouth of the Mackenzie River, at about 135° W longitude. There the party divided. Franklin led one group westward for 400 miles (644 km) as far as an inlet he named Prudhoe Bay, while the second group traveled eastward to the Coppermine River.

In the years that followed, other explorers extended Franklin's surveys and in 1845 Franklin returned with two ships, the *Erebus* and the *Terror.* Both were sailing ships with auxiliary steam engines and their bows were strengthened with iron to combat the ice. Franklin's mission was to use the surveys he and others had made and to locate the Northwest Passage. He had a crew of 129 Royal Navy personnel, and sufficient food to last three years. Unfortunately, the supplies included 8,000 cans of food that had been sealed with lead. The dangers of lead poisoning were not known then, but the metal would have contaminated the contents, making the men weak, irritable, and indecisive. Poisoning may have contributed to the expedition's failure. The two ships sailed from England on May 19, 1845. The captain of a whaling ship recorded seeing them moored to an iceberg in Lancaster Sound, to the north of Baffin Island, on July 26. They were never heard of again.

In 1847 the first search parties began looking for Franklin and his men. In all there were 39 searches. Robert John Le Mesurier McClure (or M'Clure, 1807–73), an Irish naval officer, led one of these. In 1850, McClure, in command of the *Investigator,* entered the Bering Strait from the Pacific and sailed eastward, following the coast of Alaska. He found two routes around Banks Island; one is along what is now called M'Clure Strait. The *Investigator* became trapped in ice just to the north of Banks Island and the ship had to be abandoned. Two ships from nearby Melville Island rescued McClure and his crew, but these ships were also abandoned and the party proceeded overland to Beechey Island, a small island off the south coast of Devon Island to which it is linked by a gravel causeway at low tide. From there, McClure and his crew returned to Britain by sea. On his return McClure received a knighthood and a payment of £5,000.

McClure had become the first person to travel through the Northwest Passage (albeit partly overland), but he found no trace of the Franklin expedition. Franklin's second wife, Lady Jane Franklin, spent all of her own fortune financing searches and when the British government refused to finance any further expeditions she bought a luxury yacht, the *Fox,* and had it fitted out for Arctic conditions. The first information about what had happened to the Franklin expedition came in 1854 from an Inuit hunter, who said they had perished near King William Island. The British naval officer Francis Leopold McClintock (1819–1907) took part in four searches on behalf of Lady Jane Franklin. By the time he began his fourth search, McClintock had become highly proficient at planning and executing long journeys by sledge.

In 1858 McClintock sailed in the *Fox* to King William Island, where he met local Inuit people who told him of two ships that were crushed by the ice; they gave him knives, buttons, and other items that had once belonged to members of the expedition. An old Inuit woman told of starving men who had fallen as they walked. McClintock's party went on to find a boat containing two bodies and a *cairn*—a pile of stones left as a memorial—that contained two written messages. The first, written on May 28, 1847, described the party as being well and relaxed. The second, written one year later in the margin of the first, said that Franklin and 24 other members of the party had died and the survivors were heading south. McClintock was knighted in 1860.

Later searches completed the picture McClintock had started to paint. The *Erebus* and *Terror* spent the winter of 1845–46 on Beechey Island. They continued exploring the following spring but were trapped in ice for the winter of 1846–47, and the ice remained solid through the summer of 1847. Franklin died in June 1847, and by April of the following year 21 others had died. After three more deaths, the survivors then headed south to seek help, using two boats as sledges. All of them perished.

In 1903, the Norwegian explorer Roald Engelbregt Gravning Amundsen (1872–1928) became the first person to sail successfully through the Northwest Passage. Amundsen sailed with a crew of six on the *Gjöa,* a small *sloop*—a sailboat with a single mast and fore-and-aft sails. He sailed from east to west and spent three winters trapped in ice, but finally cleared the Canadian Archipelago on August 17, 1905. He called at Nome, Alaska, and his voyage ended at Herschel Island, in the Yukon.

FRIDTJOF NANSEN AND THE *FRAM*

Ocean currents and the prevailing winds push sea ice around the Arctic Basin, and in the 1890s the Norwegian scientist and explorer Fridtjof Nansen (1861–1930) determined to track the route the ice followed. His idea was to sail to a point in the Arctic Ocean to the north of eastern Siberia, arriving in late summer. He would then allow the ship to become locked in the ice as the sea froze; he predicted that the ice would carry the ship to the Svalbard archipelago to the north of Norway, having crossed directly across the Arctic Ocean and passing close to the North Pole.

Nansen was born on October 10, 1861, at Store-Fröen, on the outskirts of Christiania (now Oslo). He studied zoology at the Royal Frederick University (now the University of Oslo). Always fascinated by Arctic exploration, in 1882 he joined the crew of the *Viking,* a sealing ship that sailed close to Greenland. A keen skier, Nansen was spurred by this glimpse of Greenland to cross the country on skis; he achieved this feat in 1888. In 1896 Nansen was made professor of zoology at the Royal Frederick University; in 1908, at his own request, he exchanged this position to become professor of oceanography. Nansen was also politically active and played a part in the dissolution from Sweden that led to Norway becoming an independent country on June 7, 1905. The photograph shown of Nansen here was taken in 1911, shortly before he embarked on his diplomatic career. Following his country's independence, Nansen was the first Norwegian minister in London and he headed the Norwegian delegation to the United States during World War I. After the war he led the Norwegian delegation to the League of Nations and helped arrange the repatriation of almost 430,000 German and Austro-Hungarian prisoners of war from Russia. In 1921 he worked with the International Committee of the Red Cross to bring relief to Russia, then suffering from famine, and he devised a scheme to issue identification documents to refugees. The Nobel Committee recognized his humanitarian work by awarding him the 1922 Nobel Peace Prize. He donated the prize money to furthering international relief. Nansen died at Lysaker, near Oslo, Norway, on May 13, 1930.

Norwegian explorer Fridtjof Nansen (1861–1930) was also a scientist and diplomat who won the 1922 Nobel Peace Prize for his humanitarian work on behalf of the League of Nations and the International Committee of the Red Cross. This photograph was taken in 1911. *(Topical Press Agency/Getty Images)*

Nansen's plan to track the Arctic Ocean currents required funding and it called for a ship capable of surviving the sea ice. The Norwegian Parliament provided part of the finance and the remainder came from private donations, including one from King Oscar II. Nansen designed the ship himself and called it the *Fram*—Norwegian for "forward." When the sea freezes around a ship

The *Fram,* shown here in about 1895, was designed and commissioned by Fridtjof Nansen. The ship was designed to rise upward as the sea froze around it rather than being crushed by the ice, so that Nansen could allow it to drift with the Arctic sea ice, thereby tracking the way the ice moved. *(Three Lions/Stringer)*

the ice exerts tremendous pressure. To escape this, Nansen gave the ship a rounded hull that would allow it to rise upward as the sea froze around and beneath it, so it rode at the ice surface without being crushed.

The *Fram,* with a crew of 13, sailed from Kristiania (the spelling had been changed) on June 24, 1893. On September 22 it had reached a position to the northeast of the Novosibirskye Ostrova (New Siberian Islands), at 78.83° N, 133.62° E. That is where the *Fram* became locked in the ice and began to drift, as Nansen had predicted it would. Nansen remained on board through the winter, expecting that the ice would carry the *Fram* close to the North Pole. It failed to do so, however, and in March 1894 he and a companion, Fredrik Hjalmar Johansen (1867–1913), left the ship to head north by dogsled and kayak. On April 8 they reached latitude 86.23° N, which at that time was the highest northerly latitude anyone had reached. They started back but had to spend the winter in a hut of moss and stone with a roof of walrus skin that they built themselves. They survived on a diet of walrus blubber and meat from polar bears and returned the following spring, arriving back in Norway on August 13, 1896. The photograph on page 163 of the *Fram* was taken in about 1895.

Nansen had been correct in calculating that Arctic sea ice drifts with the winds and currents, but mistaken about the route it followed. When it became evident that the *Fram* was not heading toward the North Pole, Nansen reasoned that this was due to the Earth's rotation, which deflected the currents.

ROBERT PEARY AT THE NORTH POLE

Nansen failed to reach the North Pole. The explorer usually credited with that achievement was Robert Edwin Peary (1856–1920), an officer in the United States Navy. Peary, accompanied by four Inuit men and Matthew Alexander Henson (1866–1955), claimed to have reached the North Pole on April 6, 1909. Peary received the formal thanks of Congress on March 30, 1911, and retired from the navy the same year with the rank of rear admiral. He died in Washington, D.C., on February 20, 1920, and was buried in Arlington National Cemetery. Henson, who was Peary's companion on most of his expeditions, died in New York City on March 9, 1955, and was buried in New York; but on April 6, 1988, his remains were reinterred at the Arlington National Cemetery, near those of Peary.

Peary entered the navy in 1881 and spent his entire career as a naval officer. The navy granted him leave to undertake his polar explorations. The first of these was in 1886, when Peary and Henson traveled over part of the Greenland ice sheet. They returned to Greenland in 1891, with a party that included Peary's wife and a physician, Frederick Albert Cook (1865–1940). This expedition proved for the first time that Greenland was an island.

In 1893–94, traveling by sledge from northeastern Greenland, Peary made his first attempt to reach the North Pole. Between 1898 and 1902 he investigated possible routes to the pole from Greenland and from Ellesmere Island, Canada. He made a second attempt on the pole in 1905, sailing the *Roosevelt,* a ship built to his specifications, to Ellesmere Island, but was able to travel no farther than latitude 87.1° N owing to bad weather and unsuitable ice for his dogsleds.

Peary left New York on board the *Roosevelt* on July 6, 1908, for his third attempt, which was successful. His party left Ellesmere Island between February 28 and March 1, 1909, and arrived at the pole on April 6.

Peary's claim to have been the first person to reach the pole, however, has always been controversial. He returned to the United States to find that Cook, his former companion, claimed to have reached the pole independently in 1908. This detracted from Peary's triumph, although Cook's claim was quickly discredited. More recently, experts examining records that had not been available at the time suggested Peary might have made navigational errors that resulted in him believing he was at the pole when in fact he was between 30 and 60 miles (48–96 km) away. Whether or not Peary was the first person to stand at the North Pole has never been resolved.

The Southern Land

The philosophers of ancient Greece believed that the universe was constructed symmetrically and that all its components had to be in balance. As geographers succeeded in mapping the known world, they assumed that the southern continents would occupy a total area similar to that of the northern continents. The principles of symmetry required it. More than 1,500 years passed before explorers charted the continents of Africa, South America, and Australia. They sailed all of the oceans, but the continents they revealed were far smaller than those of the Northern Hemisphere. Symmetry demanded a continent, so far undiscovered, to supply the missing land area. So, elusive though it was, explorers refused to abandon the search for it.

This chapter records the search for this mysterious southern continent. It tells of the early attempts that succeeded in defining the location of that continent, but without setting foot on its shore. Antarctica, the southern continent, was finally discovered early in the 19th century. The chapter describes a few of the great Antarctic explorers of the 19th and early 20th centuries, after whom some of the regions of Antarctica are named.

TERRA AUSTRALIS INCOGNITA

Until Antarctica was discovered it could have no name. Early geographers believed the philosophers who told them that the mathematical principles governing the world predicted that the total mass of land south of the equator must equal the mass of land that was known to exist north of the equator. So the geographers called this mysterious but necessary land Terra Australis Incognita, the Unknown Southern Land, and it appears on many maps drawn in the 16th and 17th centuries despite the fact that the mapmakers had no data on which to base their depictions.

Christopher Columbus used a map based on one drawn in the first century C.E. by Ptolemy (see "The Voyages of Christopher Columbus" on pages 147–152), one of the world's most famous geographers. Ptolemy did not depict the southern continent, but he certainly believed in it. His maps suggested that Africa might extend all the way to the South Pole and that there was land to the south of the Indian Ocean. Because the continent had to exist, a number of navigators believed they had found it. In 1576, the Spanish explorer Juan Fernández (ca. 1536–ca. 1604) sailed south from Chile and claimed to have found it. In 1606 the Spanish sailor Luis Váz de Torres (his dates are not known) discovered the strait between New Guinea and Australia that now bears his name while sailing as second-in-command on an expedition searching for the Terra Australis Incognita. The Portuguese leader of the expedition, Pedro Fernandes de Queirós (1563–1615) sailed through the Torres Strait, saw a large landmass to the south—which was actually Australia—and declared he had found the southern continent.

In 1642, however, Abel Tasman (see "Abel Janszoon Tasman, who Discovered Tasmania and New Zealand" on pages 141–142) sailed around Australia, proving that it was not the southern continent. Ferdinand Magellan (see "Ferdinand Magellan, From Atlantic to Pacific" on pages 132–135) had sailed around the southern tip of South America on his 1521–22 expedition, proving that this was not the southern continent. In 1488, the Portuguese explorer Bartolomeu Dias (also called Bartholomew Diaz, ca. 1451–1500) had sailed around the Cape of Good Hope, proving that Africa did not extend all the way to the South Pole, and in 1497–98 another Portuguese navigator, Vasco da Gama (ca. 1460 or 1469–1524) had sailed around the Cape of Good Hope and continued all the way to India.

Explorers had demonstrated conclusively that open ocean lay beyond the southern extremities of South America, Africa, Australia, and New Zealand, yet belief in the southern continent refused to die.

More expeditions sailed in search of it during the 18th century. All of them failed to find it.

JAMES WEDDELL AND THE WEDDELL SEA

James Cook was the first explorer to cross the Antarctic Circle (see "James Cook and Scientific Exploration" on pages 142–145), on January 17, 1773, during his second voyage in search of the fabled southern continent. Cook never found the continent, but he remained convinced that it must exist, somewhere to the south of the frozen sea.

Cook had reached latitude 71.17° S; half a century was to pass before anyone sailed closer to the South Pole. That person was James Weddell (1787–1834), an English navigator in the Southern Ocean on a sealing expedition. On February 20, 1823, Weddell sailed to 74.25° S and celebrated by hoisting a flag, firing a gun, and awarding his crew an extra ration of rum. Weddell reported seeing a few icebergs, but had no difficulty sailing between them. He saw no sign of land, however, and concluded that the continent did not exist.

Weddell was born in Ostend, Belgium, on August 24, 1787, and went to sea as a very young man. He sailed on merchant ships and served in the Royal Navy before being given command of the brig *Jane*. A *brig* is a two-masted, square-rigged sailing ship, with a gaff sail on the mainmast; a *gaff sail* is a four-cornered fore-and-aft sail. The *Jane* belonged to a shipbuilder in Leith, Scotland, and a London insurance broker, and it had been equipped for hunting seals. Weddell suggested there might be good sealing grounds in southern waters and his first voyage, to the Falkland Islands in 1819, was highly successful. His second voyage, to the South Shetland Islands in 1821–22, was less profitable, because many other sealers had headed to the region and the seals were already becoming rare, so Weddell explored farther afield. On this voyage the cutter *Beaufoy* accompanied the *Jane* and they sighted the South Orkney Islands. A *cutter* is a small, single-masted sailing ship with fore-and-aft sails, two or more *headsails*—sails carried forward of the mast—and often a bowsprit. The *Jane* and *Beaufoy* made another expedition from 1822 to 1824. They returned to the South Orkney Islands, but when they could find no seals between there and the South Shetlands, they headed south. There was little sea ice and the weather was calm, and it was on this trip that both ships sailed farther south than any other vessel before them. Weddell was not very far from the shore of Coat's Land, on the Antarctic mainland, but he decided to turn back. They named sea area that the ships had entered the Sea of George the Fourth; it is now known as the Weddell Sea.

Weddell arrived back in Britain in July 1824 and the *Beaufoy* in 1826. Weddell was then living in Edinburgh. Apparently he had a dispute with the owners of the two ships that ended with a bank suing Weddell for repayment of a loan of £245. He applied to join the navy, but was refused, and in 1829 he was still commanding the *Jane,* trading in the Atlantic. The ship was in such poor condition, however, that it had to be abandoned in the Azores. Weddell later found employment as a ship's master, sailing to Australia and Tasmania. He died in London on September 9, 1834.

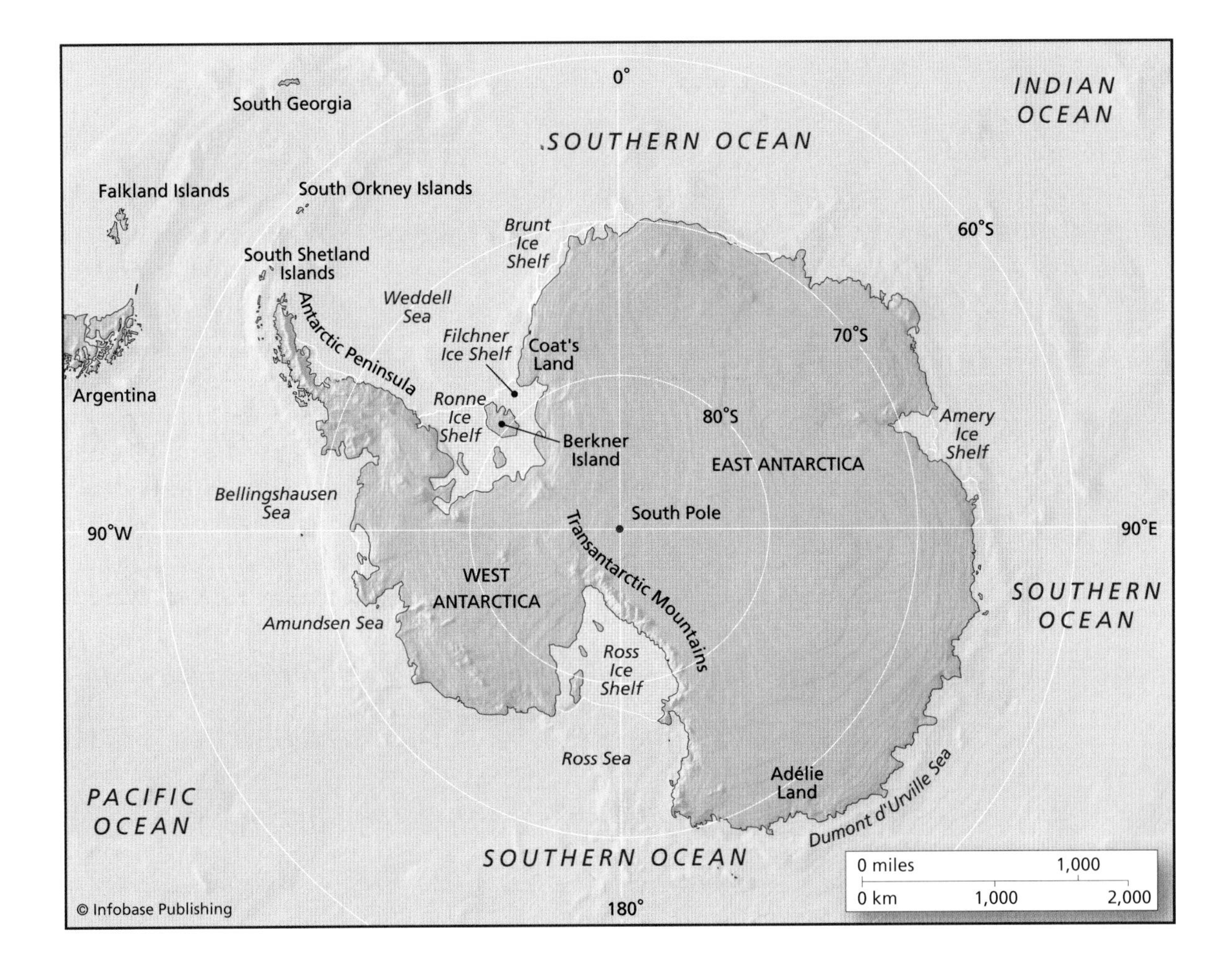

The continent of Antarctica, showing the Peninsula, Transantarctic Mountains, and principal ice shelves

It was not until 1911 that Weddell's record was broken, when the German explorer Wilhelm Filchner (1877–1957), on board the *Deutschland,* entered the Weddell Sea and discovered what Filchner called the Wilhelm II Ice Shelf; it is now called the Filchner Ice Shelf. In 1947–48 Captain Finn Ronne (1899–1980), a Norwegian-American explorer serving in the United States Navy, discovered the Ronne Ice Shelf, separated from the Filchner Ice Shelf by Berkner Island, named after the American physicist and engineer Lloyd Viel Berkner (1905–67), who took part in the 1928–30 Byrd Antarctic Expedition. The map of Antarctica on page 169 shows the location of the Weddell Sea and the largest ice shelves.

JULES-SÉBASTIEN-CÉSAR DUMONT D'URVILLE, ADÉLIE LAND—AND PENGUINS

James Weddell had reached 74.25° S in 1823. It is not often that sailors are promised a cash reward if they can break a world record, but in 1837 the carefully selected crew of the French naval ships *Astrolabe* and *Zélée* were offered 100 gold francs each if they could sail to 75° S, and an additional 20 francs for every degree they reached farther south than that. France had joined the United States and Great Britain in the scramble to explore Antarctica—and to claim areas of it.

The commander of the French expedition was Captain Jules-Sébastien-César Dumont d'Urville (1790–1842), a career officer and seasoned explorer. In 1819, while surveying in the eastern Mediterranean and Black Seas, d'Urville had learned of a recently discovered statue of great beauty on the island of Milos. He urged the French government to purchase it, the authorities agreed and instructed him to do so, and d'Urville bought the Venus de Milo. The statue is now in the Louvre Museum, Paris, and d'Urville was promoted and awarded the Légion d'Honneur.

The two ships sailed from Toulon on September 7, 1837, but by December 10 they were still north of the Magellan Strait. There had been problems with drunken brawls among the crew that had resulted in several being arrested in Tenerife, a sick officer had to be put ashore in Rio de Janeiro, and poor provisions, including rotten meat, had an adverse effect on health. D'Urville believed there was time to spend up to three weeks exploring the strait, and the sailors made use of the time catching fish and geese. They left the strait on

January 8, 1838, heading south along the coast of Tierra del Fuego, and after a few days saw their first sea ice. On the night of January 21–22 they encountered a solid wall of ice stretching across their path from horizon to horizon and were forced to turn north. After spending a few days in the South Orkney Islands the ships headed south once more. They met another ice field, but this time found a way into it. Unfortunately, the wind changed direction and ice blocked the open channel around them, trapping both ships. The men shifted ice floes by attaching ropes and dragging them, broke them with picks and axes, and eventually freed the vessels. They headed to the South Shetland Islands and discovered land, which they named Terre de Louis-Philippe; it is now called Graham Land. The expedition left Antarctica at the end of February and sailed to Chile, where d'Urville established a hospital for crew members suffering from scurvy.

The following year d'Urville determined to sail south once more, this time hoping to reach the South Magnetic Pole, at 140° E. The two ships had spent most of their time since leaving Chile visiting various Pacific islands, and on January 1, 1840, the new expedition sailed from Hobart, Tasmania. They crossed the Antarctic Circle on January 19 and sighted land the same day. On January 21 several members of the crew went ashore on an island, where they raised the French flag. D'Urville called the land beyond the island Adélie Land, after his wife. Adélie penguins *(Pygoscelis adeliae),* found all around the coast of Antarctica, are also named after Adélie d'Urville.

A few days later, while still following the coast, they saw the American schooner *Porpoise,* commanded by Charles Wilkes (see "James Clerk Ross, Charles Wilkes, and the Ross Sea" on page 172), but Wilkes sailed away from them and disappeared into the fog. D'Urville spent eight more months exploring in Antarctic waters before returning to Hobart, and heading from there to New Zealand, Guinea, Timor, and St. Helena, arriving in Toulon on November 6, 1840. D'Urville was immediately made a rear admiral and received the Gold Medallion of the Geographic Society. The other senior officers were also rewarded and the French government shared 15,000 gold francs among the 130 survivors of the two crews.

D'Urville was born on May 23, 1790, at Condé-sur-Noireau, Normandy, the son of a provincial judge. He entered the Naval Academy in Brest in 1807, a studious and serious young man, and graduated top of his class in 1811. He was interested in botany, entomology, geology, and astronomy, and he could speak Latin, Greek, English, German, Italian, Spanish, Russian, Chinese, and Hebrew. Later he learned many Polynesian languages. In 1816 he married Adélie Pepin, a Toulon clockmaker's daughter. Years of bad diet and privation weakened d'Urville's health. It was not his time at sea that killed him, however, but France's first rail crash. On May 8, 1842, he and his wife, with their son, took the train from Versailles to Paris. The locomotive derailed, overturning the carriages, with the passengers locked inside them as was the practice at the time, and the wreck caught fire. The entire family perished in the fire.

JAMES CLARK ROSS, CHARLES WILKES, AND THE ROSS SEA

On June 1, 1831, James Clark Ross (1800–62) located the North Magnetic Pole while taking part in an expedition led by his uncle John Ross (1777–1856). A specialist in the Earth's magnetic field, James Ross worked from 1835 to 1838 on a magnetic survey of Great Britain. In 1839 he sailed for Antarctica, partly in an attempt to reach the position where the German mathematician Karl Friedrich Gauss (1777–1855) had calculated the South Magnetic Pole should be.

Ross was born in London on April 15, 1800 and entered the Royal Navy at the age of 11. He sailed on his first Arctic expedition, led by his uncle, in 1818, searching for the Northwest Passage. By 1827 Ross had taken part in four more expeditions to the Arctic, so when he prepared for his Antarctic voyage he had a wealth of experience on which to draw. He took command of HMS *Erebus* and his friend Francis Rawdon Moira Crozier (1796–1848) took command of the slightly smaller HMS *Terror*. Both were sturdy, three-masted, square-rigged ships that had formerly been used to transport mortars and they were strengthened to prepare them for the Antarctic ice. They loaded ample supplies of preserved meat, vegetables, soups, cranberries, pickles, and other food items. Ross was determined that his crew should not suffer from scurvy, and he believed that a well-fed crew was a happy and hardworking crew. The sailors were given three months' pay in advance. Admiralty officials inspected the ships on September 12, 1839 and declared themselves satisfied.

They sailed on October 5 with orders to establish magnetic observing stations at St. Helena and the Cape of Good Hope before proceeding to Tasmania, where they arrived on August 15, 1840, after spending two months at the Kerguelan Islands in the Indian Ocean, where the crew took hourly magnetic readings and Ross made astronomical and tidal observations. They left Hobart on November 20 and a week later they reached the Auckland Islands, where on Campbell Island they found two boards mounted on tall poles. One recorded the visit of Charles Wilkes on March 10, 1840 and the other, the visit of Dumont d'Urville on March 11. Ross and Crozier left Campbell Island on December 17. On January 1, 1841, they crossed the Antarctic Circle and a few days later the ships encountered bad weather and thick sea ice. On January 5 Ross decided to try to force their way through the ice and early on January 9 they entered open water. That area is now called the Ross Sea. On January 11 they discovered land, which they named Victoria Land, with a range of mountains they called the Admiralty Range. Ross calculated that they were then about 500 miles (800 km) from the South Magnetic Pole. They continued westward through the Ross Sea and on January 22 they discovered an active volcano, which they called Mount Erebus, and an extinct volcano to the east of it, which they called Mount Terror.

They then headed south, only to find their way barred by a solid wall of ice rising 150 to 200 feet (45–60 m) above the sea. They called it the Victoria Barrier; the name was later changed to the Ross Ice Shelf. The ships sailed parallel to its edge for about 200 miles (320 km), but it was impenetrable. Ross decided to return to Hobart.

The *Erebus* and *Terror* headed south again on November 23, 1841. They entered the pack ice and by the end of February 1842 they were within sight of the Ross Ice Shelf, in intense cold. Once again Ross failed to find any way to penetrate the ice. They reached latitude 71.5° S before turning north and heading for the Falkland Islands. The ships made a third attempt in 1842, leaving the Falklands on December 17, hoping to sail into the Weddell Sea, but they met impenetrable pack ice and had to abandon the expedition. They reached Britain on September 2, 1843.

Ross was elected a fellow of the Royal Society in 1848. The same year he headed north in a fruitless search for Sir John Franklin's expedition (see "Franklin, McClure, and the Discovery of the Northwest Passage" on pages 159–161). Ross was knighted and reached the rank of rear admiral. He died at Aylesbury on April 3, 1862.

Charles Wilkes was born on April 3, 1798, in New York City and was educated mainly at home before commencing his naval career when he was 17. He joined the United States Navy in 1818 and in 1833, as a lieutenant, conducted a survey of Narragansett Bay, Rhode Island. This led to his appointment in 1838 as head of a government expedition to explore and survey the Southern Ocean. What was officially called the United States Exploring Expedition soon became known as the Wilkes Expedition. It included scientists, artists, and a linguist, sailing on the USS *Vincennes, Peacock,* and *Porpoise,* together with the *Relief* carrying stores, and two schooners, the *Sea Gull* and *Flying Fish.* A *schooner* is a sailing ship with fore-and-aft sails and at least two masts, the foremast usually smaller than the other masts.

The expedition sailed from Hampton Roads on August 18, 1838, and called at the Madeira Islands, Rio de Janeiro, Tierra del Fuego, Samoa, and New South Wales before sailing south from Sydney and entering the Southern Ocean in December 1839. Wilkes reported discovering a southern continent, then headed north to Fiji and Hawaii in 1840, explored the North American coast, then called at the Philippines, Borneo, Singapore, Polynesia, and the Cape of Good Hope, arriving back in New York on June 12, 1842. Wilkes rose to the rank of rear admiral. He died in Washington, D.C., on February 8, 1877, and in 1909 his remains were removed to Arlington National Cemetery.

ERNEST SHACKLETON, THE EPIC HERO

Although territorial acquisition and scientific inquiry inspired most of the early exploration of Antarctica, by the latter years of the 19th century explorers were being drawn there by the physical challenges the continent presented. In particular, there was a race to be first to stand at the South Pole. After Roald Amundsen won that race in 1911 (see "Roald Amundsen, First to the South Pole" on pages 180–182), the goal was to be first to cross the continent from coast to coast. That was the aim of the 1914–17 Imperial Trans-Antarctic Expedition led by Sir Ernest Shackleton (1874–1922).

Ernest Shackleton was born on February 15, 1874, in County Kildare, Ireland, one of the 10 children of a landowner who abandoned farming in 1880 to study medicine in Dublin. In 1884 the family moved to London, which is where Ernest spent most of his childhood. A governess taught Shackleton until he entered a preparatory school at age 11. When he was 13 he entered Dulwich College. He was not a brilliant scholar, and he left school at 16 to go to sea as an apprentice on board a square-rigged sailing ship. In 1894 he qualified as a second mate and obtained the post of third officer on a *tramp steamer*—a merchant ship that does not sail to a regular schedule visiting particular ports, but obtains cargoes wherever it can. Shackleton qualified as a master mariner in 1898. This authorized him to command any British ship anywhere in the world. In the same year he joined the Union Castle Line, carrying passengers and mail regularly between Southampton and Cape Town.

In 1901 Shackleton obtained a position on the 1901–04 National Antarctic Expedition, also known as the *Discovery* Expedition after the name of the expedition ship, and was given a commission in the Royal Naval Reserve. Commander Robert Falcon Scott (see "Robert Falcon Scott and Polar Tragedy" on pages 177–180) was leader of the expedition. The *Discovery* reached the Ross Ice Shelf on January 8, 1902. Shackleton, Dr. Edward Wilson (1872–1912), and the expedition geologist Hartley Travers Ferrar (1879–1932) traveled by dogsled from the base at McMurdo Sound, establishing a safe route onto the ice shelf. In November 1902 Scott, Shackleton, and Wilson walked to latitude 82.28° S, which they reached on December 31, but Shackleton was taken ill on the return journey and Scott decided he should return to Britain.

Shackleton's health recovered and he was able to raise sufficient financial backing to organize and lead the 1907–09 British Antarctic Expedition, or *Nimrod* Expedition, with the aim of reaching both the geographic and magnetic South Poles. The *Nimrod* sailed from New Zealand on January 1, 1908 and to save fuel the steamer *Koonya* towed it the 1,650 miles (2,655 km) from New Zealand to the edge of the sea ice. They reached open water in the eastern part of the Ross Ice Shelf where there were many whales. They called it the Bay of Whales, but the ice conditions compelled them to establish their base at McMurdo Sound. On October 19, 1908, Shackleton and three companions began their trek south. They discovered the Beardmore Glacier, named after the Scottish industrialist William Beardsmore, Shackleton's former employer, but then turned back, although three of the team, Edgeworth David (1858–1934), Douglas Mawson (1882–1958), and Alistair Mackay (1878–1914), reached the approximate position of the South Magnetic Pole on January 16, 1909. The team also made the first ascent of Mount Erebus. On their return Shackleton was knighted and received several other honors.

Amundsen had reached the South Pole in 1911, so Shackleton's second expedition as leader, the Imperial Trans-Antarctic Expedition, aimed to cross the continent from the Weddell Sea to McMurdo Sound, passing the South Pole along the way. They used two ships. The *Endurance* was a *barquentine*— a three-masted sailing ship that is square rigged on the foremast and fore-and-aft rigged on the other masts. It would carry Shackleton and the main party to the Weddell Sea and the *Aurora,* a steam yacht commanded by Aeneas Mackintosh (1879–1916), would carry a support team to McMurdo Sound. The support team would then establish depots of food and fuel across the Ross Ice Shelf as far as the Beardmore Glacier.

The *Endurance* departed from South Georgia on December 5, 1914, but encountered thick ice in the Weddell Sea. On January 19, 1915, the ship became trapped. The ice began to break in September, but in doing so it exerted extreme pressure on the ship's hull and on October 24 the *Endurance* began leaking. Shackleton ordered the ship to be abandoned at 69.08° S, 51.50° W and on November 21 the *Endurance* sank. For two months the crew camped on the ice, and made several attempts to walk to a known supply store. On April 9 they set off in their three lifeboats to reach the uninhabited Elephant Island, in the South Shetlands. From there Shackleton decided to sail to South Georgia to seek help at one of the whaling stations. Using the strongest of the lifeboats, reinforced and with raised sides, Shackleton and five companions, with supplies for four weeks, left Elephant Island on April 24, 1916. They reached the southern coast of South Georgia, 800 miles (1,287 km) distant, on May 9. After resting, Shackleton and two of the party walked across the island to the whaling stations on the northern side, from where they immediately sent a boat to rescue the three men they had left in the south. Then Shackleton persuaded the Chilean government to allow them use of a naval seagoing tug to rescue the 22 men on Elephant Island. The tug reached them on August 30. Meanwhile, the *Aurora* had carried out its mission, but had then been blown from its anchorage and out to sea. It returned to New Zealand. Three men were lost from the *Aurora,* but Shackleton's skill and leadership ensured the survival of all the others.

Shackleton arrived back in England in May 1917 and volunteered for the army. He was sent on several diplomatic missions, and spent time lecturing. In 1920 he began planning a fresh expedition, this time to sail around Antarctica. The Shackleton–Rowett Expedition, financed by his schoolfriend and now wealthy businessman John Quiller Rowett (1874–1924), departed on September 24, 1921 in a converted whaling ship, the *Quest.*

Shackleton suffered a suspected heart attack in Rio de Janeiro but refused medical help, and on January 4, 1922, the *Quest* reached South Georgia. Early the following morning Shackleton suffered a fatal heart attack. At his wife's request, he was buried at Grytviken, South Georgia.

ROBERT FALCON SCOTT AND POLAR TRAGEDY

Robert Falcon Scott (1868–1912) was a naval officer whose career began as a cadet at the age of 13. At the time of the *Discovery* Expedition he was a lieutenant with the position of torpedo officer, but his immediate family were in serious financial difficulties and he was their main economic support. He needed advancement and when he learned that the Royal Geographic Society (RGS) was planning an Antarctic expedition he visited the home of Sir Clements Markham, whom he had met some years earlier and who had since become president of the RGS, and volunteered to lead it. Scott had no Arctic or Antarctic experience or training, but plenty of enthusiasm. Markham chose him, and Scott was promoted to the rank of commander. In October 1900 Scott and Markham visited Nansen (see "Fridtjof Nansen and the *Fram*" on pages 161–164) for advice.

The *Discovery,* a sailing ship with auxiliary engines, sailed for Antarctica on July 31, 1901 with a party of 50 men. They sighted Antarctica on January 8, 1902. Soon after arriving Scott and Shackleton surveyed the area from a captive balloon. They were the first ever to do so, but the balloon developed a leak and was never used again. The party spent the winter at McMurdo Sound and began their exploration early in September. After establishing their latitude record on December 31, when they walked 300 miles (480 km) farther south than anyone had before them, the team spent a second winter at McMurdo. The following year Scott led a party inland and onto the plateau that occupies most of East Antarctica—and includes the

South Pole. As well as exploring, the team did serious scientific work in Antarctica. When the expedition ended, explosives and help from two other ships were needed to free the *Discovery* from the ice, but everyone returned safely, arriving in England on September 10, 1904. Scott was promoted to the rank of captain. The government sold the *Discovery* to the Hudson's Bay Company.

Scott returned to naval duties but was soon planning a return to Antarctica. In March 1909 he learned that Shackleton had planted a flag about 97 miles (156 km) from the South Pole. Peary reached the North Pole in April (see "Robert Peary at the North Pole" on pages 164–165). On September 13, 1909, Scott announced his plan to lead an expedition to the South Pole. The expedition also had serious scientific objectives, and although it was a private venture the British government supported it and contributed half the costs. In January 1910 Scott bought the *Terra Nova,* a three-masted, square-rigged converted whaling and sealing ship. When Scott advertised for 65 men to take part in the expedition, 8,000 applied.

They sailed from Cardiff on July 15, 1910, and Scott, who was still raising funds, joined the *Terra Nova* in South Africa and left to continue fund-raising when they reached Australia, which is where they received a telegram from Amundsen informing them that he was heading south. The expedition had become a race to the pole. Scott rejoined the ship in New Zealand. They departed on November 26 and arrived off the Antarctic coast on January 4, 1911, establishing their camp at Cape Evans, beside McMurdo Sound. The photograph on page 179 shows Scott in the base hut at that time. In addition to all the warm clothing, note the large number of books he had taken with him.

The party spent their first season leaving supplies at a series of places between the camp and a point at latitude 80° S, using dogsleds and ponies, but they were delayed starting, encountered difficult conditions, and they had to leave the final supply store at 79.48° S. Six of their eight ponies died. The team spent the winter on scientific tasks and planning for the main journey south.

That journey began on October 24, 1911, when four men left with two motor sledges to carry supplies to 80.5° S, where they would wait for the others. Both engines failed and the men had to haul the sledges themselves. The main party joined them on November 21. They were trapped for several days by a blizzard, and when it ended the surviv-

Robert Falcon Scott (1868–1912), pipe in hand, writing in his diary in his room in the expedition hut at Cape Evans, Antarctica, in January 1911. *(Time Life Pictures/Stringer)*

ing ponies had to be shot. The men continued southward, crossing the Beardmore Glacier, and on January 4, 1912, Scott announced that he and five men—H. R. Bowers, Edward Wilson, L. E. G. Oates, and Edgar Evans—would complete the final stage to the pole. On January 9 they passed the point Shackleton had reached and on January 17 they arrived at the South Pole, only to find that Amundsen had arrived there on December 14, 1911. Scott planted a British flag and the next day the party began to head back.

At first all went well, but hauling their sledges across coarse, granular snow was heavy work, and Edgar Evans was very sick. He died on February 17, near the foot of the Beardmore Glacier. All of them were suffering from malnutrition. Their pace slowed, so it took them longer to walk the average 65 miles (105 km) from one supply store to the next. Oates was suffering from severe frostbite and the effect of an old war wound. He was slowing them down, and on about March 17 he walked out of the tent to his death, saying, "I am just going outside and I may be some time." Scott, Wilson, and Bowers continued until March 20, when they were trapped in their tent by a blizzard that lasted for nine days. They had encountered weather that was unusually severe, even for Antarctica, and although every day they tried to leave their tent, they found it impossible. Scott recorded events in his diary. His final entry, on March 29, 1912, read:

> Every day we have been ready to start for our depot *11 miles* away, but outside the door of the tent it remains a scene of whirling drift. I do not think we can hope for any better things now. We shall stick it out to the end, but we are getting weaker, of course, and the end cannot be far. It seems a pity but I do not think I can write more. R. Scott. For God's sake look after our people.

On November 12, 1912, a search party from the Cape Evans camp found the tent containing the frozen bodies of Scott, Bowers, and Wilson. They later found Oates's sleeping bag, but his body was never discovered.

ROALD AMUNDSEN, FIRST TO THE SOUTH POLE

The first person to reach the South Pole was the Norwegian explorer Roald Amundsen. He was also the first person to travel to both the North and the South Pole, the first person to sail through the Northwest Passage, and one of the first people to fly across the Arctic.

Roald Engelbregt Gravning Amundsen was born on July 16, 1872, at Borge, to the south of Oslo, into a family of ship owners and sea captains. His mother did not want him to follow the family tradition and to please her he studied medicine, but she died when he was 21, at which point he promptly left university and went to sea, intent on the life of a polar explorer. His Antarctic experience began when he sailed as first mate on the steamer *Belgica* as a member of the 1897–98 Belgian Antarctic Expedition led by Adrien-Victor-Joseph de Gerlache de Gomery (1866–1934), a Belgian naval officer. The *Belgica* was trapped in ice off the Antarctic Peninsula and so this became the first expedition to spend the winter in Antarctica. In 1903 Amundsen, with a crew of six, sailed his small sloop, the *Gjöa,* through the Northwest Passage, reaching Herschel Island, Yukon, in 1905.

Amundsen's next plan was to repeat Nansen's experiment (see "Fridtjof Nansen and the *Fram*" on pages 161–164) by drifting across the North Pole in the *Fram,* thereby becoming the first person to reach the North Pole. When he learned that Peary had reached the pole in September 1909 he secretly changed his plan but continued his preparations nevertheless, and sailed from Norway on August 9, 1910. Scott had left England eight weeks earlier.

Amundsen was heavily in debt and needed a major achievement to attract funds for his projects, but dared not reveal his true plans for fear that the resulting widespread discussion might lead to them being cancelled. He told only his brother Leon, whom he could trust, and Thorvald Nilsen, captain of the *Fram.* The day before they sailed Amundsen told two more of the ship's officers of their true destination. The *Fram* sailed to Madeira, from where most people assumed it would continue to Buenos Aires, conducting oceanographic research before turning northward and heading for the Arctic. On September 9, about three hours before they sailed from Madeira, Amundsen called the crew together and told them that he intended to land a party in Antarctica and attempt to reach the South Pole. He asked each of them in turn whether they were willing to accompany him. Some went ashore. Among them was Leon Amundsen, who mailed the letters the crew had written home, and who sent a telegram to await Scott in Melbourne. This read: "Beg leave inform you proceeding Antarctic. Amundsen."

On January 14, 1911, the *Fram* reached the Ross Ice Shelf and Amundsen established his winter base camp two miles (3.2 km) inland from the Bay of Whales, choosing the location in part because the abundance of seals and penguins offered a regular supply of fresh meat. The first sign of approaching spring appeared on August 24, but it was not until September 8 that the weather was clear enough for eight men and sledges pulled by 86 dogs to depart to set up their first supply depot, at 80° S, but deteriorating weather and extreme cold forced them to unload their supplies at the depot and return to the base camp rather than continue south. On their return a disagreement with Fredrik Hjalmar Johansen (1867–1913), one of the team leaders, led Amundsen to change his plans. Johansen would lead a party exploring King Edward VII Land, while Amundsen led the attempt to reach the pole.

That attempt began on October 20. Amundsen, Olav Bjaaland (1873–1961), Helmer Julius Hanssen (1870–1956), Sverre Hassel (1876–1928), and Oscar Wisting (1871–1936) took four sledges, each drawn by 13 dogs, and arrived at the 80° S depot on October 24. They passed the point Shackleton had reached on December 8, arriving at the South Pole at 3 P.M. on December 14. The Amundsen–Scott scientific research station now stands at the South Pole. It accommodates

These workers are entering the Amundsen–Scott South Pole station, a scientific facility located at the geographic South Pole. It is named in honor of Roald Amundsen and Robert Falcon Scott. The station was built and is maintained by the United States, but accommodates researchers from many nations. *(John Beatty/Science Photo Library)*

researchers and support staff from many nations, although the station was built and is maintained by the United States. The photograph above shows workers entering the station.

Amundsen's team arrived back at their base camp on January 25, 1912, and formally announced their achievement when they reached Hobart, Tasmania, on March 7. In 1918 Amundsen sailed through the Northeast Passage, aiming to freeze his ship, the *Maud,* into the ice and allow it to drift, but this plan failed. In 1923 and 1925 Amundsen took part in attempts to fly across the North Pole, succeeding in 1926 in the airship *Norge,* which carried 15 persons in addition to Amundsen.

On June 18, 1928, Amundsen disappeared while flying in a French Latham 47 flying boat searching for Umberto Nobile (1885–1978), the Italian designer of the *Norge,* whose new airship, the *Italia,* had crashed while returning from the North Pole. The aircraft apparently crashed into the Barents Sea in fog. Amundsen is presumed to have died in the crash or soon afterward. His body was never found.

The Deserted Places

Many explorers felt drawn to the Arctic and Antarctic. Today there are research stations in these regions, staffed by scientists and support personnel, but the harsh environment continues to attract adventurers seeking to brave the challenges of traveling by sledge across ice fields, glaciers, and mountains. Others were captivated by a very different type of wilderness and ventured across the deserts of Africa, Arabia, and Asia—the deserted places. This chapter tells some of their stories.

The chapter begins with the attempts by European explorers to find Tombouctou (Timbuktu), a legendary city that became a magnet to the romantically inclined. It tells of journeys across deserts and encounters with those to whom the desert is home. The chapter ends with the discovery and exploration of that ancient trade route the Silk Road.

DIOGO GOMES, WHO MET MEN FROM TIMBUKTU

In about 1456 the Portuguese prince Henry the Navigator (see "Prince Henry the Navigator and the African Coast" on pages 119–123) sent Diogo Gomes (ca. 1420–ca. 1485) on a voyage of exploration to the coast of West Africa. Gomes set out with three caravels on a mission to learn as much as he could about the trade being conducted between African cities and to investigate the main rivers, in order to discover how far inland it was possible for ships to sail.

As well as being an experienced navigator Gomes was a diplomat who aimed to establish peaceful and friendly relations with those he met, based on mutual understanding. In 1484 Gomes dictated an account of his experiences to the German geographer Martin Behaim (see "Sir Francis Drake and the Drake Passage" on pages 136–141), who recorded it in Latin.

Gomes and his party reached the Cape Verde Islands and sailed from there toward the African coast, where they saw the mouth of the River Gambia. They entered the river with the tide, spent the night on a small island, and then continued upstream until they encountered people with whom they could trade. They exchanged cloth, necklaces, and other goods for gold, and Gomes hired a local guide and interpreter.

Leaving two of the ships behind, Gomes and the guide sailed to Cantor, a large town beside the river (now called Kuantor, Gambia). That was as far as the ship could travel, so Gomes sent his guide ahead to tell people that he was coming to trade with them. The news drew people from far and wide, among them men from Timbuktu and from a large city he called Kukia—there is no modern town of that name, and no one knows where Kukia might have been. Gomes questioned them about their rulers, trade routes, and the goods they traded. His men were suffering from the heat, however, so Gomes returned to his ship and sailed downstream to join the other two caravels. Prince Henry had sent Jacob, an Indian, to accompany Gomes and to act as interpreter should the expedition reach India. Before leaving the Gambia, Gomes sent Jacob to explore an area on the route to Timbuktu. Gomes made friends with local rulers who had previously attacked Europeans and demonstrated that he and his companions intended them no harm.

Gomes made a second journey in about 1460 as far as the Cape Verde Islands. In 1466 he was made a judge in the town of Sintra.

ALEXANDER GORDON LAING, THE FIRST EUROPEAN TO SEE TIMBUKTU

Timbuktu remained nothing but a dream in the hearts of European adventurers until 1826, when a British army officer became the first westerner to set foot in the city. Alexander Gordon Laing was born in Edinburgh on December 27, 1793, the son and grandson of schoolteachers. His father taught him until he entered the University of

Edinburgh at the age of 13 to study humanities. Two years later he began to work as an assistant teacher, but in 1809 he went to Barbados as a clerk to his uncle, Colonel (later General) Gabriel Gordon. While there Laing obtained a commission in the army, and in 1822 he was made a captain in the Royal African Colonial Corps.

In 1822, while the regiment was based in Sierra Leone, Laing was sent into the territory of the Mande, or Mandingo, peoples at the request of the governor, Brigadier Charles M'Carthy (1764–1824), with the aim of establishing trade in goods and abolishing slave trading. Later that year Laing tried to discover the source of the River Niger. Though he failed to reach it, he was able to determine its approximate location.

In 1824 Laing was sent back to England to convey the news of M'Carthy's death and to report on the state of the country following the Anglo–Ashanti War of 1823–24. While he was in England, he canvassed support for a project to visit the source of the Niger and travel the 2,600 miles (4,180 km) of its length to its delta in Nigeria. His project was approved and he was directed to travel from Tripoli, Libya, to Timbuktu, Mali, and then to explore the hydrography of the Niger Basin.

Laing left England in February 1825 and reached Tripoli where, on July 14, he married Emma Warrington, the daughter of the British consul. Leaving her behind, on July 16 Laing set out for Timbuktu accompanied by Sheik Babani, who had agreed to guide him. They reached Ghadames, near the Tunisian and Algerian borders with Libya, on September 13, where they were received hospitably. They left on October 27 and by early in December they had entered country controlled by the Tuareg, who also treated them well and told them they were 35 days' journey from Timbuktu. By late January the travelers were in what is now southern Algeria, where Tuareg tribesmen attacked their caravan at night, almost killing Laing as he slept. In a letter to his father-in-law dated May 10, 1826, Laing described his 24 injuries, 18 of them severe. He listed five saber cuts on the crown of his head and three on his left temple, all of which caused fractures. A cut on his left cheek had broken the jawbone and split his ear, and he had a bad gash on the back of his neck.

Nevertheless, Laing joined another caravan and finally arrived in Timbuktu on August 18. Earlier that year the Fulani had seized control of Timbuktu from the Tuareg and Balla, and the Fula ruler was hostile, which made Laing's position insecure. He spent his time in the city searching the records, which convinced him that his calculation of the source of the Niger had been correct.

The party left Timbuktu on September 22 heading east. On the night of September 26 Arabs attacked them while they were sleeping and Laing was killed. One of the servants accompanying Laing told the British consul that his killer was Bourabouschi, the person appointed to guide Laing from Timbuktu, who had killed him and then gone to his own country. It was later suspected that Babani had planned his murder.

RENÉ-AUGUSTE CAILLÉ, THE FIRST EUROPEAN TO RETURN SAFELY FROM TIMBUKTU

The Société de Géographie, based in Paris, is the world's oldest geographical society, founded in 1821. In 1824 it offered a reward of 10,000 francs to the first European who succeeded in visiting Timbuktu and returning safely to Europe with a description of the city. Timbuktu had an increasingly glamorous reputation as a city of wonders—and riches. Laing had reached the city but failed to return. The next person to take up the challenge did not fit the conventional image of an intrepid explorer, but he succeeded.

René-Auguste Caillé or (Caillié) was born on September 19, 1799, in the small town of Mauzé-sur-le-Mignon, in west-central France. His father—a baker and an alcoholic who had spent time in prison for theft—died when the boy was 11. René-Auguste was thin and frail, and had little education, but, inspired by Daniel Defoe's story of *Robinson Crusoe*—which was loosely based on the real-life experience of Alexander Selkirk, a Scottish privateer who spent five years marooned on the uninhabited island of Juan Fernandez—Caillé longed to visit remote places. He was highly intelligent and resourceful, courageous, and very determined.

In 1815, when he was 16, Caillé joined an expedition to Senegal. By the time he was 18 he had traveled from Senegal to Guadeloupe, in the Caribbean. At about this time he read an account of travel in Africa by the Scottish explorer Mungo Park (1771–1806). Park revealed that the River Niger flowed eastward, but he had no idea where it flowed into the sea. Caillé decided he must return to Africa. His chance came in 1824, when the Société de Géographie announced its prize.

Caillé knew that a European traveling in the desert stood a good chance of being killed, so he began his adventure by making extensive preparations. In 1824 he settled in a community near Boké, Guinea, where he learned Arabic, studied Islam, and familiarized himself with local customs. It was not until 1827 that he felt ready to join a caravan and begin the journey to Timbuktu. He wore Arab dress, carried only a bag of goods to trade, a compass, medical kit, umbrella, and journal, and he explained his poor Arabic by saying he was an Egyptian who had been taken prisoner by French troops during Napoleon's Egyptian campaign. They had taken him to France, where he had been raised, and now he was trying to go home.

He was ill on the journey, but he reached Timbuktu and stayed there for two weeks. Caillé was not impressed with the city, which he described as being a mass of mud-built houses. To return, he disguised himself as a beggar and joined a caravan bound for Tangier, Morocco. Again, he was ill on the journey north, but he finally reached Tangier. He returned to France disguised as a sailor on a French sloop. In 1828 the Société de Géographie presented Caillé with the 10,000-franc prize. He was also given a state pension and made a member of the Légion d'Honneur.

Caillé then returned to his hometown, where for a time he was mayor. In 1830 he published an account of his travels simultaneously in Paris *(Journal d'un voyage à Temboteu et à Jenne)* and London *(Travels Through Central Africa to Timbuctoo).* He died at La Badère, Charente-Maritime, on May 17, 1838. The illness that killed him was never diagnosed, but he had probably acquired it during his travels.

HUGH CLAPPERTON AND THE EXPEDITION TO CHAD

Timbuktu, now more often called Tombouctou, is in Mali and while Laing and Caillé were exploring there, other Europeans were conducting expeditions in desert lands farther to the west. The first European to visit and describe northern Nigeria and to visit another fabled location—Lake Chad—was another Scot, Hugh Clapperton (1788–1827).

Clapperton, the son of a physician and surgeon in the town of Annan, Dumfries-shire, was apprenticed as a cabin boy at the age of 13 and some years later joined the Royal Navy, either by enlisting or being press-ganged. He was promoted to officer rank and by 1817 he was a lieutenant. There was no posting for him, however, so he was placed on half-pay and sent home. In 1822 Clapperton joined the government-backed expedition to Lake Chad, together with Major Dixon Denham (1786–1828), an English army officer who had fought at the Battle of Waterloo, and Dr. Walter Oudney (1790–1824), a naval surgeon from Edinburgh. An earlier British expedition had concluded that the River Niger flowed into Lake Chad and that a river from Lake Chad flowed into the Nile. The aim of the 1822 expedition was to reach Lake Chad and check that theory.

The three men left Tripoli and crossed the desert by caravan. All three fell sick with malaria, but they persevered and reached the lake 11 months later, in February 1823. They explored the lake, but could find no river flowing into it from the west that was large enough to be the Niger and no major river flowing out of it on the eastern side.

The team then separated. Denham headed southeast, following the River Chari upstream toward its source in what is now the Central African Republic. Clapperton and Oudney went west searching for the Niger, and joined a caravan that was going to the city of Kano in what is now Northern Nigeria. Oudney died in January 1824, but Clapperton continued to Kano and went from there to Sokoto, which in those days was a much more important city. Sokoto then lay inside the Fulani empire, established by a formerly nomadic people who had adopted a settled way of life, and the ruler of Sokoto was Mohammed Bello. Bello was friendly toward Clapperton and interested in establishing good relations with Britain, but he refused to allow Clapperton to leave and tried to discourage him from exploring the Niger. Eventually he relented and provided an escort to take Clapperton eastward, with a letter to the British king expressing his desire to trade with Britain and his willingness to collaborate in ending the slave trade.

Clapperton reached Kukawa, near the shore of Lake Chad, where he met Denham, who had been searching for him. They headed north together in September 1824, reaching Tripoli in late January 1825. There, Clapperton learned that Laing had been authorized to explore the Niger. When they reached London the British authorities welcomed the letter from Mohammed Bello, and Clapperton found himself a hero.

Later that year Clapperton returned to Africa, landing on the Nigerian coast in November. He made for Sokoto, but three of his four English companions soon died from malaria, leaving only one, his manservant Richard Lander. On the way to Sokoto, Clapperton saw the Niger for the first time. When they reached Sokoto Clapperton found that Bello was involved in a war and was no longer interested in cooperating with the British. In February 1827 Bello allowed Clapperton and Lander to leave by sailing down the Niger. On the journey Clapperton fell ill with malaria and dysentery. Despite dedicated nursing from Lander, he died on April 13, 1827.

JAMES RICHARDSON, HEINRICH BARTH, AND ADOLF OVERWEG IN NORTH AND CENTRAL AFRICA

Between 1850 and 1855 the Prussian geographer Heinrich Barth (1821–65) made one of the most comprehensive studies of North and central Africa. Barth was born in Hamburg on February 16, 1821, and studied classics at the University of Berlin. He was fluent in Arabic—and also spoke as French, Spanish, Italian, and English—and had already explored the North African coast before embarking on a journey across the Sahara.

The 1850 expedition in which he participated left from Tripoli. It was sponsored by the British government and led by the English explorer James Richardson (1806–51). Its aim was to open commercial relations with the nations of central Africa. Richardson, born in Boston, Lincolnshire, on November 3, 1809, had earned his reputation traveling in India, but he was a seasoned traveler in the Sahara. He had spent nine months in 1845 journeying from Tripoli to Ghadames and Ghat, Libya, collecting information about the Tuareg people. The team also included the German geologist and astronomer Adolf Overweg (1822–52). Overweg was to become the first European to travel completely around Lake Chad and to sail on its waters.

The party traveled in a southwesterly direction, toward what is now Northern Nigeria. Along the way the group separated, Overweg proceeding alone through Zinder, in what is now southern Niger, and then eastward to Kukawa, in northeastern Nigeria, where he joined the others. Richardson died close to there on March 4, 1851, and Barth took command. He and Overweg traveled southward and around the southern shore of Lake Chad. Overweg fell ill and died at Maduari, Chad, on September 27, 1852. Although by then his own health was poor, Barth continued alone, traveling westward and eventually reaching Timbuktu. He stayed there for six months—longer than any other European had—before heading back to Tripoli and from there to London.

By the time his journey ended, Barth had traveled about 10,000 miles (16,000 km). He was able to describe the middle section of the River Niger and he had recorded the routes he had followed using dead reckoning. Between 1857 and 1858 he published an account of his travels, together with details of the peoples he met, their histories, and their languages. The resulting four volumes, *Reisen und Entdeckungen in Nord- und Central-Afrika in den Jahren 1849 bis 1855* (Travels and discoveries in North and Central Africa in the years 1849 to 1855), provide one of the fullest descriptions of those parts of Africa ever written. In 1863 Barth was appointed professor of geography at the University of Berlin. He died in Berlin on November 25, 1865.

HENRI DUVEYRIER AND THE TUAREG

After his travels through the Sahara, Heinrich Barth returned to London where he spent some time writing his account of the expedition. While he was there he met Henri Duveyrier (1840–92), a young Frenchman with an ambition to follow in his footsteps who was living in London at the time. To prepare for his journey, Duveyrier learned Arabic. He was only 19 years old when he traveled to North Africa for the first time, in 1859. He spent nearly three years there, and on his return to Paris in 1863 the Société de Géographie presented him with its Gold Medal. In 1864, still only 24 years old, Duveyrier published *Exploration du Sahara: Les Touâreg du nord* (Exploration of the Sahara: The Tuareg of the north). This was the first of his journeys through North Africa. His later travels took him to the region immediately to the south of the Atlas Mountains and to the shallow salt lakes of Algeria and Tunisia. Duveyrier died in Sèvres on April 25, 1892.

Duveyrier was especially fascinated by the Tuareg. He lived among them for months at a time and familiarized himself with their speech and customs. Known as the Blue People because the men wear blue robes and turbans, the Tuareg were once farmers leading a settled life, but in the 12th century an invasion by Bedouin Arabs drove them from their lands and they became nomadic pastoralists. Today there are rather more than 5 million Tuareg living in Niger, Mali, Algeria,

Burkina Faso, and Libya. They speak a language called Tamasheq, which is a dialect of the language spoken across much of the Sahara by non-Arab peoples.

In the 19th century, when Duveyrier met them, the Tuareg were feared as bandits. Armed with daggers, lances, shields, and double-edged swords, bands of these blue-robed warriors riding horses or camels would appear as though from nowhere to rob travelers, then vanish as mysteriously as they came. Seen from the Tuareg point of view, however, this was just business. The Tuareg were traders who also organized caravans to transport merchandise across the desert. Travelers paid to join their caravans and the Tuareg taxed caravans that were not under their control. As recently as the 1940s, the Tuareg owned and operated approximately 30,000 caravans. They ran the truck companies of the desert, but today real trucks have largely replaced camel caravans and many of the Tuareg have either become farmers or have settled in the cities, although wars have left a number living in refugee camps around Timbuktu.

FRIEDRICH GERHARD ROHLFS, CROSSING THE DESERT

Most of those who spent time in the Sahara were on diplomatic missions or engaged in scientific research, but as well as these explorers there were also adventurers. Gerhard Rohlfs was an adventurer. He was born on April 14, 1831, in Vegesack, now a district of Bremen, Germany, the son of a physician. He was educated in a gymnasium (high school) at Osnabrück, but when he was 17 he enlisted in the army and saw active service before studying medicine at the universities of Heidelberg, Würzburg, and Göttingen.

The desert exerted a powerful attraction, however, and in 1855 Rohlfs went to Algeria to join the French Foreign Legion. He was decorated for bravery and during his time in the Legion he learned Arabic and studied local customs. In 1862, disguised as an Arab, Rohlfs explored Morocco and the Atlas Mountains, and then entered the heart of the Sahara, penetrating as far south as the Fezzan region of Libya, but his guides attacked and robbed him, leaving him for dead. He managed to make his way back to Algeria and made two more excursions into the desert before spending a few months in Germany.

In March 1865 Rohlfs returned to Tripoli, intending to explore the Ahaggar highlands of southern Algeria. The Tuareg were at war, however, making the region unsafe, so he went instead to Ghadames. From there he traveled to the western side of Lake Chad, then followed the Benue River to its confluence with the Niger, and finally followed the Niger to its delta on the Gulf of Guinea. Rohlfs had become the first European to cross the Sahara from north to south. He traveled from the Niger Delta to Lagos and sailed from there to England, arriving in Liverpool on July 2, 1867.

In 1868 Rohlfs traveled from Tripoli to Egypt. Then he returned home to Germany, married, and settled down in the city of Weimar. He returned to Africa in 1873 to lead an expedition into the Libyan Desert and he took part in further expeditions in 1878 and 1880.

Rohlfs was made German consul in Zanzibar in 1865. There was keen rivalry in East Africa between the British and Germans, and the German government was keen to secure the island of Zanzibar. Rohlfs had no diplomatic training, however, and was soon recalled to Germany. He never visited Africa again. Rohlfs died at Rungsdorf, near Bonn, on June 2, 1896.

CARSTEN NIEBUHR AND THE ARABIAN COAST

King Frederick V of Denmark was the first European ruler to send a scientific expedition to study the antiquities of Arabia, Egypt, and Syria. It was led by a German surveyor, Carsten Niebuhr (1733–1815). The team of five set out in 1762. They visited the Nile, crossed it into Sinai, and then sailed southward along the Arabian coast as far as Jidda. From Jidda they traveled overland to Mocha (al-Mukhā) in the far southwest. The philologist (a philologist is a person who studies the language and literature of a people) who was traveling with the group died in May 1763, and the naturalist died in July. The three survivors—an artist, a surgeon, and Niebuhr—visited Ṣan'ā', in Yemen, then returned to Mocha. Niebuhr wrote two books about the region. *Beschreibung von Arabien* (Description of Arabia) was published in 1772 and *Reisebeschreibung nach Arabien und andern umliegenden Ländern* (Description of travels to Arabia and other surrounding countries) was published in 1774. For many years these were regarded as classics.

Carsten Niebuhr was born on March 17, 1733, in Lüdingworth, now a district of Cuxhaven, Lower Saxony. His father was a farmer and Carsten had little formal education, but he enjoyed mathematics and geography, and was able to study surveying at the University of Göttingen. He was still studying surveying when one of his teachers suggested he apply to join the expedition the Danish king was planning. Niebuhr redoubled his efforts to master the intricacies of surveying and to improve his mathematical skills, and he also learned some Arabic. He was a talented linguist and his adoption of local dress and food probably helped him remain healthy in the hot, dry climate.

The Danish expedition arrived back in Copenhagen in November 1767 and Niebuhr was given a position in the Danish military, which entitled him to live in Copenhagen. He married in 1773. In 1778 he accepted a civil service position in Holstein, Germany. He died at Meldorf, Schleswig-Holstein, on April 26, 1815.

CHARLES MONTAGU DOUGHTY AND THE INTERIOR OF ARABIA

Two volumes of a different kind of travel book appeared in 1888. *Travels in Arabia Deserta* was written by the English traveler Charles Montagu Doughty (1843–1926). At first the book aroused little interest, but T. E. Lawrence (popularly known as Lawrence of Arabia, 1888–1935) discovered it and arranged for it to be republished in 1920, with an introduction that he wrote for the reissue. From that point, the work's popularity grew over the years, until it came to be seen as a masterpiece of travel writing.

Doughty was born at Theberton Hall, at Leiston, Suffolk, the son of a clergyman. He attended private schools and studied geology at King's College London and Gonville and Caius College, University of Cambridge, where he graduated in 1864. But it was not the rocks of Arabia that he wished to examine. He was seeking out ancient inscriptions and searching for what he imagined was the birthplace of human culture. He believed the desert Bedouin—the nomadic peoples of the Sahara, Arabian, and Middle Eastern deserts—closely resembled the earliest humans and he lived among them. He dressed simply, but he never attempted to pass himself off as an Arab or tried to hide the fact that he was English and a Christian.

In 1876 Doughty set out from Damascus, traveling with hajj pilgrims bound for Mecca. He did not reach Mecca, but he visited Madā' in Sālih, near Medina, and on his return journey he called at Taymā', Hā'il, 'Unayzah, and at-Tā'if, towns in the mountainous region of Jabal Shammar, deep inside Arabia, as well as the port of Jidda. He returned to England in 1878 and died on January 20, 1926, at Sissinghurst, Kent.

After its rediscovery by Lawrence, *Travels in Arabia Deserta* was valued as much for its literary style as for its content. It is truly a masterpiece. The following excerpt, describing the scene in Damascus, where Doughty was waiting to join the hajj, gives a flavor of the work.

> There is every year a new stirring of this goodly Oriental city in the days before the Haj; so many strangers are passing in the bazaars, of outlandish speech and clothing from far provinces. The more part are of Asia Minor, many of them bearing over-great white turbans that might weigh more than their heads: the most are poor folk of a solemn countenance, which wander in the streets seeking the bakers' stalls, and I saw that many of the Damascenes could answer them in their own language. The town is moved in the departure of the great Pilgrimage of the Religion and again at the home-coming, which is made a public spectacle; almost every Moslem household has some one of their kindred in the caravan. In the markets there is much taking up in haste of wares for the road. . . . Already there come by the streets, passing daily forth, the *akkâms* with the swagging litters mounted high upon the tall pilgrim-camels. They are the Haj caravan drivers, and upon the silent great shuffle-footed beasts, they hold insolently their path through the narrow bazaars; commonly ferocious young men, whose mouths are full of horrible cursings: and whoso is not of this stomach, him they think unmeet for the road. . . . The assembling of the pilgrim multitude is always by the lake of Muzeyrîb in the high steppes beyond Jordan, two journeys from Damascus. Here the hajjies who have taken the field are encamped, and lie a week or ten days in the desert before their long voyage.

Doughty adopted a literary style that was almost Elizabethan, and he seemed more intent on producing a masterpiece than a travel guide. The result, however, is a description written in language that distances the reader and conveys an impression of scenes that are alien and remote.

SIR WILFRED THESIGER, WITH THE BEDOUIN AND THE MARSH ARABS

In more recent times the British soldier, traveler, photographer, and travel writer Sir Wilfred Thesiger (1910–2003) kept the romance of the desert alive. Thesiger was born on June 3, 1910, in Addis Ababa, Ethiopia, and he lived there until he was nine. He could, therefore, claim northern Africa as his home. He was educated at Eton College and studied history at Magdalen College, University of Oxford. He returned to Africa in 1930, having received a personal invitation from Emperor Haile Selassie to attend his coronation. His explorations in Africa commenced in 1933, when he traced the course of the Awash River in Ethiopia. Thesiger served in the British army with the Special Air Service (SAS), holding the rank of major, and he fought in Ethiopia and North Africa during World War II. After the war he worked with the Desert Locusts Research Organization in Arabia.

Thesiger made several crossings of the Rub' al-Khali—the Empty Quarter of the Arabian Desert—but his main interest was the Ma'dan, or Marsh Arabs, with whom he lived for some time. They live in southern Iraq in a region of marshland between the Euphrates and Tigris. The marshes form a triangle with the towns of An-Nāṣirīyah, Al-'Amārah, and Basra at its corners. Thesiger described them in his book *The Marsh Arabs,* published in 1964.

During his travels through Arabia Thesiger lived with the desert nomads, the Bedu or Bedouin. He described their way of life in his book *Arabian Sands,* published in 1959, and deeply regretted the steady erosion of their traditions. Sir Wilfred Patrick Thesiger died at his home near Croydon, Surrey, on August 24, 2003.

FERDINAND VON RICHTHOFEN, WHO DISCOVERED THE SILK ROAD

European explorers were also venturing farther east, some of them on scientific missions. One of the most influential scientists was the German geologist and geographer Ferdinand, Freiherr (Baron) von Richthofen (1833–1905). Von Richthofen participated in several expeditions to Asia and in the 1870s traveled extensively in China. He published a description of that country in a five-volume work with an atlas, *China, Ergebnisse eigener Reisen and darauf gegründeter*

Studien (China, the results of my own travels and studies based on them). This work appeared between 1877 and 1912, and in it von Richthofen referred to what he was the first to call the *Seidenstraßen* (Silk Roads), a term he used to describe the old trade routes between China and Europe (see "The Silk Road" on pages 59–61).

Von Richthofen was born on May 5, 1833, in what was then the town of Carlsruhe, in Prussian Silesia (it is now called Pokój, in western Poland). He studied geology at the Universities of Breslau (now Wroclaw) and Berlin and made his scientific reputation with studies of the Dolomite Alps (northern Italy) and the geology of Transylvania. He first visited Asia from 1860 to 1862 as a member of a Prussian diplomatic expedition aiming to promote trade with China, Japan, and Thailand. The expedition visited Sri Lanka, Japan, Taiwan, Indonesia, the Philippines, Thailand, and Myanmar (Burma). When the mission ended, von Richthofen went to the United States, where he worked as a geologist from 1863 until 1868—and discovered gold in California. He was professor of geology at the University of Bonn from 1875 to 1883, when he became professor of geography at the University of Leipzig, and in 1886 he was made professor of geography at the Friedrich Wilhelm University of Berlin (now the Humboldt University of Berlin). He died in Berlin on October 6, 1905.

SVEN ANDERS HEDIN ON THE SILK ROAD

One of von Richthofen's students continued the exploration of the Silk Road. In 1893 the Swedish explorer Sven Anders Hedin (1865–1952) began a five-year journey across the Ural and Pamir Mountains, past Lop Nor, and to Beijing, along the old Silk Road. Later Hedin discovered the way Lop Nor had formed, from shifts in the course of the Tamir River. While he was there the area had no inhabitants. The groups of Uighur people who moved into the region later left around 1920, because of a plague that killed many of them. The lake itself, which covered about 770 square miles (2,000 km^2) in 1950, ceased to exist in 1970, when dams held back the water of the Tamir that used to feed into it. Between 1899 and 1902 Hedin explored the Gobi Desert and from 1927 to 1933 he led a Sino-Swedish expedition that discovered evidence of Stone Age cultures in what is now desert.

Hedin was born in Stockholm on February 19, 1865, and was educated first in Uppsala and from 1881 to 1883 in Berlin and Halle.

It was while he was in Germany that he fell under the influence of von Richthofen and decided to devote his life to exploration. He wrote many books describing his travels. These included *Through Asia* (1898), *Scientific Results of a Journey in Central Asia,* published between 1904 and 1907 as 10 volumes of text and two of maps, *Southern Tibet,* published between 1917 and 1922 as 11 volumes of text and three of maps, *The Silk Road* (1938), and his autobiography *My Life as an Explorer* (1926).

Hedin's travels began in 1885, with a journey through the Caucasus, Iran, and Mesopotamia (now Iraq) that was made possible when he secured a position as tutor to a family in Baku, the capital of Azerbaijan and a port on the west coast of the Caspian Sea. In 1890 he served as an interpreter on a Swedish–Norwegian mission to Iran and was able to travel eastward as far as the Chinese border.

The time he spent in Germany made him a supporter of that nation during World War I and he had close contacts with the Nazi leaders during the 1930s and 1940s. Although he supported certain Nazi policies, however, he was strongly critical of others, in particular of the persecution of the Jewish people and the suppression of academic and religious freedom, and he was instrumental in securing the release of a number of people from concentration camps and under sentence of death. Hedin died in Stockholm on November 26, 1952.

SIR AUREL STEIN AND THE CAVES OF A THOUSAND BUDDHAS

The name most closely associated with the rediscovery of the Silk Road, however, is that of Hungarian-born Mark Aurel Stein. Stein was born in Budapest on November 26, 1862, became a British citizen in 1904, was knighted in 1912, and died in Kabul, Afghanistan, on October 26, 1943.

Sir Aurel Stein conducted three expeditions, in 1900, 1906, and 1913, into what is now the Xinjiang Uygur Autonomous Region of China. These lasted for a total of seven years and followed the old caravan routes between China and the west for about 25,000 miles (40,200 km). Stein traced the Silk Road across Lop Nor, and found the Jade Gate that had once stood at the border between China and the Western Kingdoms and the walls that had been built to keep nomads from entering China. As he went, Stein excavated ancient sites carefully, discovering long-lost cities and, near the city of Dunhuang (Tun-Huang), the Cave of the Thousand Buddhas.

The Cave of the Thousand Buddhas is a temple that was carved into a cliff in 366 C.E., at the start of a period lasting until the early 13th century, when the city was a major Buddhist center. Eventually there were nearly 500 temple caves, now known as the Mogao Caves. Many are now open to the public, and in 1987 the area was declared a World Heritage Site. In one of the caves Stein discovered a hoard of about 60,000 paper manuscripts and other documents, dating from the fifth to 11th centuries, that had been walled up in 1015. They included Buddhist, Taoist, Zoroastrian, and Nestorian scriptures, stories and ballads, and were written in Chinese, Sanskrit, Tibetan, Uighur, and other languages. There were also paintings and temple banners. Stein removed many of them, and the French scholar Paul Pelliot (1878–1945) later removed more. These returned to London and Paris, where they were properly stored, cataloged, and published, but news of their removal reached the Chinese authorities in Beijing. They bought the remaining material and dispatched it to Beijing, but the money to pay for it was stolen before it reached Dunhuang and most of the material was stolen before it reached Beijing. Much of it has now been lost.

Other Worlds

Some time between 1620 and 1630, the German mathematician and astronomer Johannes Kepler (1571–1630) wrote a fantasy entitled *Somnium* (The dream), in which mysterious forces carry a student all the way to the Moon, and the student describes the Earth as it might appear to someone in space. Kepler's story may have been the first work of science fiction, but Kepler was not the first person to fantasize about space travel. It is a very ancient aspiration.

The dream of being able to fly far above the Earth, and the possible consequences of doing so, occur in Greek mythology, in the story of Daedalus and Icarus. Daedalus had offended Minos, the legendary king of Crete, and the king imprisoned him together with his son Icarus. Daedalus was smart, however. He obtained feathers and wax, with which he fashioned two pairs of wings, allowing father and son to escape. Icarus was too ambitious, however. He flew too close to the Sun. The wax holding his wings together melted and he fell to his death in the sea. Centuries later, the Greek author Lucian (ca. 120–ca. 180 C.E.), in his *True History,* satirized the beliefs of many poets and historians, and included an account of an imagined flight to the Moon.

In September 1944 the Germans began to fire ballistic missiles at targets in France, Belgium, the Netherlands, western Germany, and England. They called them V-2 (*Vergeltungswaffe*—"retaliation force") rockets and they were capable of reaching a height of more than 100 miles (160 km). That is suborbital height. The V-2s flew to the edge of space.

On October 4, 1957, the ancient dream took its next major step toward realization. That was when the Soviet Union launched *Sputnik-1,* the first artificial satellite. Weighing 184.3 pounds (83.6 kg), *Sputnik* circled the Earth in an elliptical orbit at a height ranging from 134 miles (215 km) to 583 miles (939 km), and at a speed of about 18,000 MPH (29,000 km/h). It took 96.2 minutes to complete one orbit. *Sputnik* transmitted a radio signal that allowed surface stations to track it. The signals ended when its batteries were exhausted after 22 days, but the satellite remained in orbit until January 4, 1958. It had flown approximately 37 million miles (60 million km).

Eventually 25 satellites were launched during the *Sputnik* program, although seven failed. *Sputniks* carried the first animals to fly in space. Laika, a dog, flew on *Sputnik-2* in November 1957, and other animals flew on later missions. The *Sputnik* program spurred the United States to better the Soviet achievement, but on April 12, 1961, Senior Lieutenant Yuri Alexeyevich Gagarin (1934–68), a fighter pilot in the Soviet air force, became the first human to orbit the Earth. The photograph reproduced below shows him strapped

On April 12, 1961, Yuri Gagarin (1934–68), an officer in the Soviet air force, became the first person to orbit Earth, traveling in the spacecraft *Vostok 1.* He is seen here strapped into the cockpit of his spacecraft. *(Novosti/Science Photo Library)*

into the cockpit of the spacecraft *Vostok-1.* He completed one orbit of the Earth at a height of 187.75 miles (302 km). The journey lasted 108 minutes. Gagarin, by then a colonel, was killed in a plane crash on March 27, 1968.

This chapter brings the history of exploration up to the present and hazards a gaze into the possible future. It describes the Apollo missions that carried the first humans to the Moon. It tells of the missions to Mars and the possibility of manned missions to that planet and the construction of permanent bases there. It speculates about the feasibility of travel beyond the solar system. And finally it discusses the search for evidence of intelligent life elsewhere in the universe.

THE APOLLO PROGRAM

Yuri Gagarin became the first cosmonaut in April 1961. In May of the same year, the National Aeronautics and Space Administration (NASA) announced the Apollo Program to carry an American astronaut to the surface of the Moon and return him safely to Earth. The Apollo Program ended in December 1972. In all it flew 33 missions into space, of which 11 were manned. The unmanned missions were used to test the *Saturn V* launch vehicle and all of the emergency procedures.

The first planned mission with a crew ended in catastrophe on January 27, 1967, when a fire broke out inside the *Apollo 1* spacecraft during a rehearsal for the launch, killing all three of the astronauts on board: Edward White, Virgil Grissom, and Roger Chaffee. The first successful manned mission took place on October 11, 1968. *Apollo 7* carried Walter Schirra, Jr., Donn Eisele, and Walter Cunningham into Earth orbit. They completed 163 orbits before returning to the surface.

Apollo 8 was the first mission to carry a crew to the Moon. The spacecraft was launched on a three-stage Saturn V rocket—it was the first manned mission to use the *Saturn*—on December 21, 1968, with commander Frank Borman, command module pilot James A. Lovell, Jr., and lunar module pilot William A. Anders onboard. The craft entered Earth orbit, where it remained while all its systems were checked, then it was dispatched to the Moon. It took 66 hours to reach the Moon, where the spacecraft entered lunar orbit and the crew spent 12 hours photographing and filming the surface. *Apollo 8* returned to Earth and landed in the Pacific on December 27. The Apollo 9 mission, with James A. McDivitt, David R. Scott, and Russell L. Schweickart, lasted for 10 days. It launched on March 3, 1969, and its task was to check the performance of the crew and all the equipment. The mission included extravehicular activity. *Apollo 10,* with Thomas P. Stafford, John W. Young, and Eugene A. Cernan, launched on May 18, 1969, and its mission lasted for eight days. It

NEIL ARMSTRONG

On July 21, 1969, the United States astronaut Neil Armstrong became the first person to stand on the surface of the Moon.

Neil Alden Armstrong was born on August 5, 1930, at Wapakoneta, Ohio. He enrolled in 1947 under the Holloway Plan to study aerospace engineering at Purdue University, graduating in 1955. (The Holloway Plan paid the tuition fees for those who committed to four years of study followed by three years of service in the U.S. Navy and then two final years of study.) Armstrong received a master's degree in 1970 from the University of Southern California.

Armstrong entered the navy in January 1949 to train as a pilot, qualifying in August 1950. He saw action in 1951 and 1952 during the Korean War, where he flew 78 missions. He left the navy in August 1952 and entered the Naval Reserve, and resigned from the Reserve in October 1960. Armstrong returned to Purdue in 1952.

After graduating from Purdue, Armstrong became a civilian test pilot. In 1957 he flew his first rocket-powered airplane, the Bell X-1B, and in November 1960 he flew a North American X-15 to an altitude of 48,840 feet (14.9 km), reaching a speed of Mach 1.75 (equivalent to 1,003 knots/1,153 MPH/1,855 km/h). In 1957 Armstrong was chosen for the U.S. program to put a man in space, and in 1962 he was accepted for astronaut training by the National Aeronautics and Space Administration (NASA).

On March 16, 1966, Armstrong went into space for the first time as command pilot on the Gemini 8 mission, with pilot David Scott (1932–). As part of the preparation for lunar missions, the *Gemini* crew docked with an unmanned target vehicle and Armstrong undertook extravehicular activity. On September 12, 1966, Armstrong flew as backup pilot on the Gemini 11 mission. Armstrong was backup commander on the Apollo 8 mission to orbit the Moon, but did not fly. In December 1968 he was made commander of *Apollo 11.*

After the Apollo 11 mission Armstrong declared he would not go into space again and he resigned from NASA in August 1971. He was university professor of aerospace engineering at the University of Cincinnati from 1971 until 1979. Armstrong received many honors, including the Presidential Medal of Freedom, the Congressional Space Medal of Honor, and the freedom of the burgh of Langholm, Scotland (the seat of clan Armstrong).

American astronaut Buzz Aldrin is seen here standing on the surface of the Moon's Mare Tranquillitatis (Sea of Tranquillity) on July 20, 1969, in a photograph taken by the mission commander, Neil Armstrong, whose reflection is visible in the visor on Aldrin's helmet. *(NASA)*

rehearsed undocking and docking. The Apollo 11 mission achieved the first landing on the lunar surface—it launched on July 16, 1969, carrying Neil A. Armstrong as commander (see the sidebar on page 202), Michael Collins as command module pilot, and Edwin E. (Buzz) Aldrin, Jr. as lunar module pilot. The spacecraft entered lunar orbit approximately 76 hours after it had been launched, and the lunar module, called *Eagle,* landed in Mare Tranquillitatis (Sea of Tranquility) on July 20. Armstrong and Aldrin spent the next two hours checking all the systems on the module and setting the controls for their stay on the surface. Both men then donned their portable life-support equipment and Neil Armstrong climbed out through the forward hatch and stepped down the ladder onto the surface. Aldrin followed him soon afterward. The photograph above reproduced here shows Aldrin on the lunar surface. The two astronauts spent their planned 2.5 hours on the lunar surface, during which time they collected about 46 pounds (21 kg) of surface material for examination on Earth, then reentered the module. After resting they left the surface, made rendezvous and docked with the orbiting command module, and then returned to Earth.

Apollo 12 demonstrated that it was possible to fly to the Moon and land at a precise spot on the surface in rough terrain. Charles Conrad, Jr., Richard F. Gordon, Jr., and Alan L. Bean launched on November 14, 1969, and landed in the Oceanus Procellarum (Ocean of Storms). The crew spent two periods on the lunar surface, one of four hours and the other of 3.75 hours.

Apollo 13, carrying James A. Lovell, Jr., Fred W. Haise, Jr., and John L. Swigert, Jr., launched on April 11, 1970. It reached the Moon, but 56 hours into the mission one of the two liquid oxygen tanks in the service module failed, losing the entire content. This left the spacecraft with sufficient oxygen for the fuel cells to provide basic electrical power for about two hours, and the command module returned to Earth using very little power, but the crew were able to land safely.

Apollo 14, manned by Alan B. Shepard, Jr., Stuart A. Roosa, and Edgar D. Mitchell, launched on January 31, 1971. The crew spent a total of 9.25 hours on the lunar surface. *Apollo 15,* carrying David R. Scott, Alfred M. Worden, and James B. Irwin, launched on July 26, 1971. It was the first of three missions designed to spend more time on the lunar surface and to travel greater distances using a lunar roving vehicle. Its crew spent a total of 18.6 hours on the Moon and traveled 17.3 miles (27.9 km). *Apollo 16,* with John W. Young, Thomas K. Mattingly II, and Charles M. Duke, Jr., launched on April 16, 1972. The crew spent 20.2 hours on the surface, traveled 16.6 miles (26.7 km), and collected 207 pounds (94 kg) of rock samples. *Apollo 17* was the final mission of the program. It launched on December 7, 1972, with Eugene A. Cernan, Ronald E. Evans, and Harrison H. Schmitt. They spent a little more than 22 hours on the lunar surface, collected 242 pounds (110 kg) of samples, and traveled approximately 21 miles (34 km).

MARS

Apollo ended in 1972, but new plans were being laid in 2009 for a return to the Moon, this time perhaps to establish a permanent base on the surface. The United States plans to build a permanent lunar base between 2015 and 2020. Commercial organizations were debating the feasibility of hotels on the Moon to accommodate (extremely wealthy) tourists. But behind all such schemes lies the ultimate dream of sending explorers to Mars.

It is an old dream. Probably it began in 1659. That is the year when the Dutch astronomer and physicist Christiaan Huygens (1629–95) drew the first picture of Mars as seen through his telescope. Huygens revealed that Mars has a solid surface and is a world, just as Earth is a world, and therefore a place to visit, with mountains to climb and plains to cross. Venus, Earth's other neighbor in the solar system, hides its surface beneath a thick layer of perpetual cloud. The German-born English astronomer Frederick William Herschel (1738–1822) emphasized the similarities between Earth and Mars, suggesting in 1784 that the ice caps visible at the planet's poles were made from snow and ice, just like those on Earth.

The image of Mars became somewhat distorted when, in 1877, the Italian astronomer Giovanni Virginio Schiaparelli (1835–1910) described lines on its surface and used the word *canali* to describe them. The Italian word means "channels," but it was mistranslated as "canals" and in 1894 the American astronomer Percival Lowell (1855–1916) suggested the canals were evidence of a martian civilization that employed irrigation systems on a scale so vast that they were visible from Earth. Such ideas inspired the notion of an old and dying civilization. In 1898 the British author H. G. Wells used this as the basis for his novel *War of the Worlds,* in which Martians invade Earth. Although attractive, unfortunately the idea was quite wrong. The lines that Schiaparelli saw existed only inside his telescope. By the early 20th century astronomers were convinced that the surface of Mars was extremely arid and bitterly cold, with a very thin atmosphere consisting mainly of carbon dioxide and with no oxygen. The planet was deeply hostile to the kinds of living organisms known on Earth.

That did not mean the planet should not be explored, and attempts to send spacecraft to the vicinity of Mars began in 1960 with two Soviet missions, both of which failed to launch. After several more failures, on June 19, 1963, the Soviet *Mars 1* craft (also known as *Sputnik 23*) approached to within 120,000 miles (193,000 km) of

Mars. Success finally came with NASA's *Mariner 4,* launched on November 28, 1964. *Mariner 4* flew past Mars on July 14–15, 1965, in an orbit that brought it to a height of 6,118 miles (9,846 km) above the surface. The photographs the spacecraft took of the surface were the first images of another planet ever to be returned to Earth from space.

After several more American and Soviet missions that placed spacecraft in orbit about Mars and obtained more detailed information about the surface, the most dramatic development occurred on July 20, 1976, when a lander detached from the *Viking*

The rock-strewn surface of Mars, photographed by a camera on board *Viking 1.* The image was prepared by separating the colors from the *Viking* photograph, then using a computer to paint them back onto a high-resolution black-and-white image. The computer then balanced the colors, making this image what a person standing on the planet's surface would probably see. *(Mary A. Dale-Bannister, Washington University, St. Louis)*

1 orbiter, landed safely in a smooth, circular region known as Chryse Planitia, and took photographs of its surroundings. One of those photographs is reproduced on page 206. For the first time, people on Earth saw the surface of Mars from ground level. The picture was of a rolling landscape covered with sand and scattered rocks, beneath a pale yellowish-brown sky. *Viking 2* was launched a few weeks behind *Viking 1.* It also landed safely and transmitted pictures of the surface.

One aim of the Viking missions was to seek evidence of life on Mars. The results were slightly confusing, because martian surface chemistry is unlike soil chemistry on Earth, but they revealed no conclusive evidence of living organisms. Nevertheless, that search has continued. It is important, because if living organisms exist on Mars or have ever lived there in the past, they will be the first examples of extraterrestrial life to be discovered. Such organisms would be studied intensively, and they would need careful protection from the risk of contamination by organisms transported to Mars from Earth.

Mobile vehicles—rovers—have since explored small areas of Mars, transmitting large volumes of data on the surface geology and geomorphology of the planet. The Mars Exploration Rovers *Spirit* and *Opportunity* landed on January 24, 2004, and were still returning images and data in November 2009. On October 14, *Spirit* had traveled 4.8 miles (7.73 km) across the surface of Mars, and on October 29, *Opportunity* had covered 11.57 miles (18.62 km).

Future missions aim to gather samples from the martian surface and return them to Earth. The Mars Sample Return Lander mission is planned to launch in 2014 and return to Earth in 2016, and further missions to orbit, land, and explore the surface, perhaps returning samples, are scheduled for 2016. All of these missions will be international collaborations between NASA, France, Italy, and other nations, and they will prepare for the realization of the ancient dream of landing humans on Mars. The United States is hoping to launch such a mission in 2019 or soon after that.

Traveling to Mars will not be like traveling to the Moon. The journey to the Moon takes a few days, while that to Mars takes several months. Moreover, the journey is feasible only at intervals 26 months apart, because the distance between Earth and Mars varies on that timescale from about 35 million miles (56 million km) to 249 million miles (400 million km). Once astronauts land on Mars they will have to live there for more than two years before Earth and Mars are close enough to allow them to return. They cannot carry supplies for such a long period, so while they are on Mars they will have to establish a base inside which they can produce their own food, recycle water, generate power, and provide themselves with everything else they need. Radio messages between Mars and Earth will take about 15 minutes to travel each way, so an astronaut making an urgent inquiry will have to wait at least 30 minutes for a reply.

The first human explorers on Mars will need to be as self-reliant as explorers in the wildest, most remote and inhospitable regions of Earth. Once they have established the first base settlement, however, later migrants will be able to enlarge and improve it. Little by little conditions will become easier, just as they have for the scientists who work in Antarctica. Eventually, the bases—for in time there will be several—will be occupied permanently. Perhaps, one day, people will be born on the Moon and on Mars, as the first natives of those worlds.

EXPLORING THE SOLAR SYSTEM

In 1961 the Soviet Union launched two probes to Venus. Both failed, as did the first U.S. mission, *Mariner 1,* in 1962. In August 1962 NASA's *Mariner 2* was the first spacecraft to fly close to Venus, passing within 22,000 miles (35,000 km) of the surface. On March 1, 1966, the Soviet probe *Venera 3* crashed onto the surface of Venus. This was the first craft to reach the surface of another planet. In 1967, the Soviet *Venera 4* entered the planet's atmosphere, returning data that showed this to consist mainly of carbon dioxide and nitrogen. It also revealed that the atmosphere was much denser than scientists had expected, and that the surface was extremely hot. On December 15, 1970, *Venera 7* landed on the surface of Venus and returned data from there. This was the first vehicle to land on the surface of another planet. On October 22, 1975, *Venera 9* landed on Venus, on a slope covered with boulders, and returned the first photographs ever taken on the surface of another planet. The probe found that the surface atmospheric pressure was about 9 MPa (90 times the surface air pressure on Earth), the temperature was 900°F (485°C), and that the light level beneath a yellowish sky was about equal to that of a cloudy summer day on Earth. Because of the heat and pressure, the probe survived for only 53 minutes. It is unlikely that humans will ever land on Venus!

Since then other missions have visited all the planets of the solar system and most of the many moons that orbit them. In 1984 the Soviet *Vega 1* placed a balloon into the atmosphere of Venus then flew on to intercept the comet Halley. This helped prepare the way for *Giotto,* a mission by the European Space Agency (ESA), which flew to within 370 miles (596 km) of the nucleus of Halley in March 1986.

In 1991 the *Galileo* spacecraft passed within 1,000 miles (1,600 km) of the asteroid Gaspra and photographed it, and in 1993 *Galileo* visited and photographed the asteroids Ida and its tiny moon Dactyl. As it flew past Earth, *Galileo* also studied Earth's atmosphere and radio emissions to see if it was possible to detect the presence of life from deep space. It found these, opening the possibility of detecting life from afar on other planets. *Galileo* was the first spacecraft to orbit Jupiter and to release a probe into its atmosphere. The probe was launched on October 18, 1989, and reached Jupiter on December 7, 1995.

In 2001 the NEAR–Shoemaker mission reached the asteroid Eros and on February 12 of that year the spacecraft landed on the asteroid's surface. NEAR stands for Near Earth Asteroid Rendezvous; Eugene Shoemaker (1928–97) was an American geologist who specialized in the study of comets and asteroids. Eros is approximately 21.4 × 7.0 × 7.0 miles (34.4 × 11.2 × 11.2 km) in size.

On October 15, 1997, the Cassini–Huygens spacecraft, named after the Italian-French astronomer Giovanni Domenico Cassini (1625–1712) and Christiaan Huygens, departed on a mission that would take it past Venus, Earth, and Jupiter, to Saturn and to the moons of Saturn. NASA, ESA, and the Italian Space Agency (ASI) jointly sponsored the mission, and the spacecraft entered its orbit around Saturn on July 1, 2004.

On December 25, the two parts of the spacecraft separated and the *Huygens* probe headed for Titan, the largest moon of Saturn, which was discovered by Huygens in 1610. Titan interested scientists because the *Voyager 1* probe, launched in 1977 on a mission to visit Jupiter and Saturn, found that Titan possessed a dense atmosphere and there was evidence suggesting there might be liquid at the surface. On January 14, 2005, *Huygens* arrived at Titan and after a descent through the atmosphere lasting 2.5 hours it landed on the surface and began returning pictures. The lander was standing on solid ground, but its pictures showed no sign of the ocean of liquid methane scientists had anticipated, although a photograph taken during the descent showed a landscape with hills and features resembling a shoreline and river valleys. Later images obtained by *Cassini* discovered what may be lakes, one approximately the size of Lake Superior and another the size of the Caspian Sea. The liquid in these lakes would be ethane or methane, and Titan was found to have clouds made of methane droplets, suggesting that from time to time methane rains onto the surface. No one will be visiting Titan anytime soon, however. The surface temperature is -290°F (-179.5°C)!

Titan was one of the possible objectives for the next mission to the outer planets. In the end it lost, because the planners did not feel that the mission was at a sufficiently advanced stage technically. Instead, in February 2009 NASA and ESA agreed to implement the Europa Jupiter Mission.

Europa is one of the four moons of Jupiter that were discovered in 1610 by Galileo Galilei (1564–1642). Its diameter is approximately 1,940 miles (3,121 km) and it orbits 416,900 miles (670,900 km) from Jupiter. The moon's entire surface is covered in ice, and *Galileo* confirmed that beneath the ice there is an ocean of salt water. No one knows how thick the ice is or the temperature of the water beneath it, but there is a possibility that organisms might live in the water, where they would be protected from the intense radiation from Jupiter. Europa also possesses a very thin atmosphere consisting mainly of oxygen. Clearly there is much for scientists to investigate. The mission should launch in 2020.

IS STAR TRAVEL POSSIBLE?

The spacecraft *Voyager 2* was launched on August 20, 1977, bound for Jupiter, Saturn, Uranus, and Neptune. By September 25, 2009,

Voyager 2 was about 8.4 billion miles (13.4 billion km) away from Earth.

Voyager 1 was launched on a mission to Jupiter and Saturn on September 5, 1977. Although it left Earth after its twin, it was called *Voyager 1* because it was launched onto a faster path and so reached Jupiter and Saturn first. In 2003 *Voyager 1* was close to the outermost boundary of the solar system. By September 25, 2009, it was approximately 10.36 billion miles (16.67 billion km) from Earth and traveling at about 38,200 MPH (61,470 km/h) relative to the Sun. That is about 0.006 percent of the speed of light, which travels at about 186,000 miles (300,000 km) per second. The nearest star to the Sun is Proxima Centauri, a red dwarf with no known planets, about 4.2 light-years away. At its present speed *Voyager 1* would reach Proxima Centauri in about 72,000 years. The closest star that might possess a planetary system is Epsilon Eridani, 10.8 light-years away, or 185,000 years away at *Voyager* speed.

It has taken the two *Voyager* craft more than 30 years to reach the edge of the solar system. The distances to other stars are several orders of magnitude greater, and the distances to other galaxies are many times greater than that. Clearly it is possible to launch a mission to another star, because although the *Voyagers* are not heading for any particular star they are entering interstellar space. Spacecraft launched today would be more powerful than the *Voyagers* and would fly faster and could be directed toward a target star, but by the time they reached their destination, tens of thousands of years later, there is no certainty that the civilization that sent them would still exist. Distance is the problem.

That does not make manned interstellar travel totally and forever impossible. If space voyagers abandoned any hope of arriving at another star, they could travel in ships that were entire worlds. Such vessels would provide not simply farms and workshops to supply the material necessities of life, but education to the highest levels, research facilities, theaters, concert halls, art galleries, and the artists to use them. The passengers would board the ship, probably in Earth orbit, and they would live there for the rest of their lives. Many thousands of years later their descendants would enter the gravitational field of the target star.

Alternatively, a different type of propulsion might allow a spaceship to travel at a significant fraction of light speed. Pressure exerted by a very powerful laser beam directed at a sail up to 600 miles (1,000 km) in diameter and made from highly reflective material could slowly accelerate a ship until it was flying at as much as half of 1 percent of the speed of light. The laser would be mounted in space, and once the starship attained its planned cruising speed it would coast for the remainder of its journey. As the starship approached another star it might reverse part of its sail so the starlight decelerated the craft.

A pulse engine that produced a series of controlled nuclear explosions could also power a starship. It would be costly, but it would allow the ship to attain a very high speed, perhaps up to 10 percent of light speed. At that speed, the journey to Epsilon Eridani would take about 108 years.

Sails and pulse engines could be constructed using technology that is available today. Several projects have been designed that are based on them and space agencies are watching with interest a program to test a sail that is being implemented by The Planetary Society. Other propulsion systems call for more advanced technologies. The engines cannot be built now, in the early years of the 21st century, but they may well become feasible by the middle of the century.

At present no one is planning an interstellar voyage, but one day someone will. Until then there is much more exploration to do within the solar system.

SEARCH FOR EXTRATERRESTRIAL INTELLIGENCE

Frank Drake, born in Chicago in 1930, is professor emeritus of astronomy and astrophysics at the University of California, Santa Cruz, and he has many years of experience in radio astronomy. Drake is best known, however, for having devoted much of his career to the search for life on other planets. In 1961 a number of astronomers, physicists, social scientists, and industrialists met at Green Bank, West Virginia, to discuss the likelihood that there are civilizations on other planets and that it might be possible to find evidence of them. Drake attended that meeting and as part of his preparation for it he devised a mathematical equation that might be used to estimate the number of civilizations in the Milky Way galaxy. The Drake equation, also known as the Drake formula and the Green Bank formula, is:

$$N = R^* \times f_p \times n_e \times f_l \times f_i \times f_c \times L$$

where: N is the number of civilizations, R^* is the average number of new stars that form in the Milky Way each year, f_p is the fraction of those stars that possess planets, n_e is the average number of planets capable of supporting life possessed by each star that has planets, f_l is the fraction of those planets that develop life, f_i is the fraction of those that develop intelligent life, f_c is the fraction of civilizations that reaches a stage of technological development sufficient for them to emit detectable signs of their existence, and L is the length of time during which those civilizations continue to emit detectable signs. When Drake fitted numbers to his equation the result was:

$$N = 10 \times 0.5 \times 2 \times 1 \times 0.01 \times 0.01 \times 10{,}000 = 10$$

Drake suggested that the galaxy contained 10 civilizations capable of being detected. Clearly, the solution to the equation depends on the values allotted to each of the variables, and most of these have to be chosen arbitrarily. Many astronomers consider Drake's values to be too high, but even their lower values yield a result greater than two, meaning there is a high probability that the Milky Way galaxy supports at least one advanced civilization in addition to the civilization on Earth.

People have believed in other inhabited worlds at least since the days of ancient Greece, when the philosophers Thales (ca. 624–ca. 546 B.C.E.) and Anaximander (ca. 610–ca. 546 B.C.E.) accepted that there were many such worlds. Some people believed the Moon was inhabited, as were Venus and Mars. As astronomers discovered more about the conditions on these bodies it became evident that humans could not survive on them without carrying with them their own life-support systems, and that there were no intelligent beings on those worlds with which human beings might converse. The vision of intelligent extraterrestrial beings faded.

Today the tide has turned and most astronomers believe it extremely unlikely that Earth is the only body in the galaxy where living organisms are to be found. Groups around the world are searching for other planetary systems and on March 6, 2009, the NASA Kepler Mission was launched into space to search for Earth-like planets orbiting other stars.

The active search for intelligent life began in 1960, with Project Ozma, when Frank Drake trained the radio telescope at Green Bank on the two stars Tau Ceti and Epsilon Eridani, but it detected nothing. In 1963 a much bigger radio telescope, called Big Ear, built by Ohio State University, began operation at Delaware, Ohio. Some time later Big Ear began the world's first continuous search for extraterrestrial intelligence, or SETI. In 1971 NASA funded a study for a SETI project called Project Cyclops that would have used a single radio telescope with 1,500 receiving dishes. This was never built. In 1979 the University of California, Berkeley, began SERENDIP—the Search for Extraterrestrial Radio Emissions from Nearby Developed Intelligent Populations. This program was updated several times and still continues. NASA's Microwave Observing Program (MOP) planned to survey the whole sky and to target 800 nearby stars. It was funded by the U.S. government and began in 1992, but one year later Congress withdrew its funding and MOP ended before it even began.

The SETI Institute is now the principal organization engaged in the search. It was founded in 1984 as a nonprofit corporation and it has three centers, the Center for SETI Research, the Carl Sagan Center for the Study of Life in the Universe, and the Center for Education and Public Outreach. The SETI Institute employs more than 150 scientists and other staff. Frank Drake heads the Carl Sagan Center.

So far, no signal from an extraterrestrial (ET) civilization has been detected. There have been several intriguing signals, but none has been repeated—and for an ET signal to be credible, it must be repeated. The searchers look for radio signals in a narrow band, like the signal from a radio station on Earth, but signals that cannot have originated on Earth. The receiving equipment being used is listening only for the carrier wave. If such a signal were to be received, the equipment would not be capable of receiving the signal—the message—carried on that wave, only of identifying it as an intelligent signal from a particular point in the sky. If the signal were to be confirmed by several observatories and the scientists agreed about its intelligent origin, the news would be passed to the United Nations. The UN member states would then decide whether or not a reply should be transmitted and what such a reply should contain.

The radio telescopes around the world that are engaged in SETI collect a vast quantity of data. Searching for an intelligent signal necessitates processing all of that data. That is a formidable task and calls for far more computing power than is available to the SERENDIP team at the University of California, Berkeley, but in 1999 the team found a solution. Instead of relying on one impossibly expensive supercomputer they invited members of the public to allow them to use their home computers when these were idle and would typically be running a screensaver. The scheme is called SETI@home and it was the first example of distributed computing. When participants are not using their computers but leave them switched on, the program automatically downloads a batch of data onto the home computer, processes it, and returns the results. By the summer of 2009 more than 3 million home computers were taking part in the scheme.

An alien civilization might not choose to communicate by radio, preferring an alternative: light. Given sufficient power, an intense laser pulse would be detectable over a distance of many light-years. The SETI Institute, together with the Lick Observatory belonging to the University of California, Santa Cruz, is scanning many star systems for bright pulses that arrive less than one-billionth of a second apart. There is no natural process that could emit laser pulses of this kind, so they must have been sent deliberately by an intelligent civilization and beamed at the solar system. That would mean that the presence of intelligent life on Earth had been detected.

It would not be possible to hold a conversation with an extraterrestrial civilization because of the distances involved. If such a civilization were discovered at a distance of, say, 10 light-years, then 10 years would have elapsed since the message received on Earth had been transmitted and a reply would take 10 years to travel back. It would be difficult to hold a conversation with a 20-year gap between making a remark and receiving a reply—and the likelihood is that the intelligent beings are much more than 10 light-years distant. All of that is assuming that the two civilizations, on Earth and a distant planet, were able to understand each other well enough to communicate at all.

No other intelligent civilization has yet been discovered, but the search has been running for only a few decades. It may be that no signal will ever be received, but on the other hand, the certainty that at least one extraterrestrial civilization exists would have profound implications for our own ideas of our place in the universe.

Why We Explore

Animals survive by being alert to their surroundings. They must be able to locate food, avoid predators, find mates, and protect themselves against harsh weather. They must also seek fresh opportunities. Human exploration has always been opportunistic. Explorers have traveled in search of new resources to exploit, new land to cultivate, new markets for their goods.

This final, very short chapter describes the earliest origins of those searches, the very birth of exploration. It begins with the first and longest journey of them all.

THE LONG WALK OUT OF AFRICA

The ancestors of all modern humans lived in East Africa. *Homo sapiens* is the taxonomic name for modern humans and our immediate ancestors are also placed in the genus *Homo.* The earliest member of the genus was *H. habilis,* which lived from about 2.4 million years ago until about 1.4 million years ago in South and East Africa. Other species followed and *H. sapiens* first appeared about 250,000 years ago. Humans and their immediate ancestors, as well as their close relatives the gorillas and chimpanzees, are placed in the subfamily Homininae and are known as hominins. About 20 species are known, of which nine belong in the genus *Homo.*

The hominins evolved in Africa, but gradually some of them began moving elsewhere. *H. erectus,* which lived from about 1.4

million years ago until 2 million years ago, moved from Africa into Asia, eventually reaching China and Java. *Homo floresiensis,* which was discovered in 2003–2004, was probably descended from *H. erectus.*

About 70 million years ago a number of modern humans—*H. sapiens*—living in what is now Ethiopia moved northeast until they arrived at the shores of the Red Sea. Perhaps they were following game animals, or finding plant foods that became better tasting or more abundant the farther they walked; no one knows. About 50,000 years ago the world entered its most recent ice age. As the polar ice sheets grew thicker, accumulating water taken from the oceans, sea levels fell and the strait at the southern end of the Red Sea became narrower and shallower, and perhaps dotted with islands. The humans crossed it, presumably on rafts. For some time they had been harvesting shellfish along the shore, and perhaps it was the search for shellfish that took them into Arabia.

Once arrived, their migration continued. One group of their descendants moved around the Arabian coast and eventually arrived in India. Another group traveled north and reached Europe, where they encountered another hominin species, *H. neantherthalensis.* The Neanderthals became extinct about 30,000 years ago, but no one knows whether modern humans played a part in their disappearance. Other humans moved to Central Asia and from there some walked south into India. By about 35,000 years ago there were humans living in Siberia, Korea, and Japan. From the vast continent of Eurasia people then moved across the Bering Strait from Siberia into North America (see "Crossing the Bering Strait" on pages 42–45). Farther south, other people sailed from Asia across the Pacific and eventually to the Pacific Islands and to Australia (see "Colonizing the Pacific Islands" on pages 38–41 and "Arrival in Australia" on pages 41–42).

The journey took approximately 35,000 years, so it proceeded very slowly. At each stage, the migrating groups were following signs that led them in the direction of new or richer resources. Not everyone moved, of course. A group would settle in an area that supplied them with everything they needed. Some time, or several generations, later, a number of them would leave what had become their ancestral home. They would move on, always searching, always seeking a better life. They were exploring their world and, in doing so, they were occupying and taking possession of it. It was the longest journey on Earth that humans have ever taken.

SURVIVAL AND THE NEED TO KNOW

Those early travelers, moving on foot or on rafts or in simple boats, were in no hurry. They had no defined destinations or a purpose that took them from one place to another. They were following the migrations of the animals that supplied them with food, skins, bones, horns, antlers, and other materials, walking a little farther to find plant foods when they had consumed most of those nearby. Occasionally a small party, seeking more and bigger fish or shellfish perhaps, would take to the water to reach a shore they could see in the distance. Between migrations communities would spend years—probably generations—moving around the same area along well-trodden tracks that led from one good foraging area to the next and eventually returned them to their starting point.

They learned to read the signs in the sky that told of impending changes in the weather. They understood the behavior of the animals they pursued and of the large, dangerous carnivores that pursued them. Their knowledge of the seasons allowed them to prepare for winter and to long for the return of the Sun.

People were busy finding and preparing food, making clothing, making and repairing shelters, making tools, and avoiding danger. They were surviving, and as they did so they were acquiring knowledge of the world around them. They passed on their knowledge and also the questions to which they could find no answers. Their well-being depended crucially on their skill at interpreting the world they lived in, making the best possible use of its resources, and finding new resources. Knowledge provided access to those new resources and revealed new ways to increase the utility of traditional resources. The people were highly inventive.

From the very beginning, when *Homo erectus* left Africa and moved into Asia and *Homo sapiens* crossed into Arabia, humans have been characterized by their curiosity and their resourcefulness. People are explorers by nature, and their explorations have taken them to every corner of the world, and now beyond. Humans explore with their minds as well as with their feet. Their curiosity demands explanations for the phenomena they observe, and the search for those explanations has led to the emergence of organized, methodical, rigorous research and also to the creation of works of art. As a result of all that exploration and curiosity, the human species has flourished and continues to do so.

Conclusion

People are explorers not out of a desire for thrills or the delight of new experiences, but because the need to know what lies beyond the horizon, to find explanations for the puzzling phenomena that nature generates, and to understand how the world and the universe function is deeply ingrained in human nature. With the capacity to understand there comes the ability to predict, and those who can predict reliably what will happen next tend to live longer than those who lack that talent. Physical and mental exploration helps us survive and it is a large part of what makes us human.

This book has sketched the history of human exploration as a series of brief snapshots. It has looked at some of the earliest ships, which were the vehicles that made large-scale exploration possible. It told of some of the early voyages, and it recounted the lives and exploits of some of history's most famous navigators. As well as being explorers, humans are traders, and much exploration was driven by the search for markets and for goods to sell in them. Marco Polo, perhaps the most famous of all travelers, was the son and nephew of merchants.

Not all the explorers crossed the ocean. Others traveled across continents and deserts. They visited the Arctic and Antarctic.

Finally, humans ventured into space. In 1971 the Soviet Union launched *Salyut-1,* the first space station, and in 1973 the United States launched its own space station, called *Skylab.* The first modules of the Soviet *Mir* space station were placed in orbit in 1986.

Each of these space stations was designed, built, and owned by either the Soviet Union or the United States, although there was considerable collaboration between the space agencies of the two nations. The change came with the concept of the International Space Station (ISS), which was designed and built from the start as a joint venture, with hardware contributed by the United States, Russia, the European Space Agency, Japan, Canada, and Brazil. Construction began in 1998, and in 2000 the first crew arrived. Since that time the ISS has been manned at all times.

Humans are now permanently resident in space, away from Earth. The next phase in human exploration has begun.

GLOSSARY

alluvial deposits materials transported by streams and rivers and found on riverbeds and floodplains.

aphelion the date on which Earth is at its farthest from the Sun; at present this is July 4.

aquifer a mass of permeable material such as sand or gravel that lies above a layer of impermeable material such as clay or solid rock, and through which water is able to flow.

back-quadrant *see* **backstaff.**

backstaff (back-quadrant) a navigational instrument for measuring latitude, derived from the **cross-staff,** in which the user stands with his or her back to the Sun and looks along the staff to align the shadow cast by a moveable vertical arm on a vane at the end of the staff with the horizon seen through a slit in the vane.

barquentine a three-masted sailing ship that is square rigged on the foremast and fore-and-aft rigged on the other masts.

bireme a **galley** with two rows of oars on each side.

boom a rigid support along the bottom edge of a sail.

boltsprit *see* **bowsprit.**

bowsprit (boltsprit) a pole that extends forward from the **prow** of a sailing ship, to which the stays from the fore-mast can be attached, allowing the fore-mast to be positioned farther forward than would be possible otherwise.

brig a two-masted, square-rigged sailing ship, with a **gaff sail** on the mainmast.

cairn a pile of stones left as a memorial.

caravan a group of travelers crossing a desert together.

caravanserai (khan) an inn with an inner courtyard where **caravans** rest overnight.

caravel a two- or three-masted sailing ship (later four-masted) with a **lateen rig** on all masts.

catamaran a vessel with two hulls.

chip log a ship's **log** comprising a quarter-circle piece of wood attached by a bridle to a knotted line. It was thrown from the ship's stern and remained stationary in the water while the ship moved away from it at a speed measured by paying out the knotted line.

ciguatera an acute illness producing vomiting, diarrhea, blurred vision, toothache, and other neurological symptoms that is caused by eating fish contaminated with the **dinoflagellate** *Gambierdiscus toxicus.*

clinker built made from planks the length of the vessel, with each plank overlapping the one below it.

crinoline a balcony or partial deck in a **galley.**

cross-staff (Jacob's staff) a medieval instrument, consisting of a graduated staff and moveable cross-piece, that was used to measure the angle of elevation of celestial objects and, in navigation, to calculate latitude.

cutter a small, single-masted sailing ship with fore-and-aft sails, two or more **headsails,** and often a **bowsprit.**

dead reckoning a method of navigation based on estimating position from the measured speed and direction traveled since the last known position.

declination the angle between a celestial object and the horizon.

dinoflagellate a member of a large division (Dinomastigota) of single-celled, mobile, aquatic (mainly marine) organisms.

dugout canoe a canoe made by hollowing out a single tree trunk.

equinox one of the two days each year when, everywhere on Earth, the Sun is above the horizon for exactly 12 hours and below it for 12 hours, making day and night of equal length.

fore-and-aft sail a sail that can take the wind on either side by swinging across the ship.

gaff sail a four-cornered fore-and-aft sail.

galley a seagoing ship that can be rowed.

gore one of the sections of maps that fit together on the spherical surface of a globe.

grease ice sea ice comprising a covering of ice crystals that makes the sea appear oily and that is starting to separate into flat pieces of ice.

gunwale (pronounced "gunnel"), the upper edge of the ship's side.

halyard one of the ropes used to raise and lower a **yard.**

headsail any sail carried forward of the most forward mast.

hoplite in ancient Greece, a citizen-soldier who fought with a spear and round shield.

Jacob's staff *see* **cross-staff.**

janissaries crack infantry units that formed the sultan's bodyguards and household troops in the Ottoman Empire. They were mostly conscripted Christian boys who converted to Islam.

keel the structure that extends from bow to stern along the center of a ship's bottom, adding strength and directional stability to the ship.

khan *see* **caravanserai.**

knot a speed of one **nautical mile** per hour.

Lapita culture a culture, identified by its distinctive pottery and domesticated animals and crops, that developed in the Bismarck Archipelago and spread from there to Polynesia, where it formed the basis of Polynesian culture.

larboard *see* **port.**

lateen rig a triangular sail held on a long yard with the forward end very low and the aft end high, held on a mast that is inclined forward.

latitude the angular distance from a line passing around the equator.

lee side the side sheltered from the wind.

letters patent a legal document issued by a monarch or government in the form of an open letter granting certain rights or status to an individual or entity such as a city or corporation.

lodestone (magnetite) a naturally occurring iron-oxide mineral ($Fe^{2+}Fe^{3+}{}_2O_4[FeO.Fe_2O_3]$) with magnetic properties.

log the device used to measure the speed of a ship.

log book a book containing a record of all journeys undertaken by a ship or aircraft.

longitude the angular distance from a line passing through Greenwich, England, and through the geographic North and South Poles. A line of longitude is called a meridian.

loxodrome *see* **rhumb line.**

magnetic variation the angular distance between magnetic and geographic north or south.

magnetite *see* **lodestone.**

master in the days of sail, the person in charge of the ship's navigation and steering. The master was not necessarily the ship's commander.

meridian *see* **longitude.**

mordant a substance used in dyeing to make colors fast.

nautical mile a distance based on the length of one minute of arc of latitude, standardized since 1929 as 1,852 meters (6,076.4 feet). The unit is widely used by aircraft as well as ships.

oasis a natural depression in a desert that is deep enough for its bottom to be above the level of water in an underground **aquifer,** so water lies at or close to the surface, within the reach of plant roots.

obelisk a tall stone column that tapers upward and ends in a point. The earliest sundials were obelisks.

octant a navigational instrument, based on a 45° arc, used to measure the angle between two distant objects.

outrigger canoe a small, narrow boat that is stabilized by one or two long floats fastened by struts to the main hull.

papyrus *Cyperus papyrus,* a **sedge** that grows in marshes. It was used to construct boats, mats, and paper in ancient Egypt.

patent log *see* **taffrail log.**

peltast in ancient Greece, a light infantry soldier armed with shield and javelins.

perihelion the date on which Earth is at its closest to the Sun; at present this is January 4.

periplus a set of written instructions for navigators, literally a "sailing around."

plane of the ecliptic an imaginary disk with the Sun at the center and a circumference traced out by the Earth's orbital path.

port (larboard) the left side of a ship when facing forward.

post captain in the British Royal Navy, a naval rank as opposed to a courtesy title used of anyone commanding a ship.

precession the movement of the axis of a rotating body caused by the application of a force to the axis. The axis moves in a direction 90° to the applied force, in the direction of rotation.

precession of the equinoxes the gradual westward movement of the equinoxes along the **plane of the ecliptic** due to the Earth's **precession.**

prow the part of a ship that projects forward from the **stem.**

quadrant a navigational instrument in the form of a quarter-circle (quadrant). The user held it with the upper edge horizontal, aligned a moveable arm with a celestial object, and read the angle of elevation from a scale on the curved edge.

quadrireme a **galley** with four rows of oars on each side.

quarter the part of a ship approximately one-fifth of the distance from the stern to the bow.

quinquereme a **galley** with five rows of oars on each side.

rhumb line (loxodrome) a line drawn on a map that crosses all meridians of **longitude** at the same angle.

saga an Icelandic or Norse history of a king or important family, often recounting heroic deeds or adventures.

sailing master the member of the crew of a sailing ship responsible for navigation and determining the deployment of sails.

schooner a sailing ship with fore-and-aft sails and at least two masts, the foremast usually smaller than the other masts.

sea loch in Scotland, a deep inlet that is open to the sea.

sedge a flowering plant belonging to the family Cyperaceae, found in wetland habitats, that resembles grass and rush.

sextant a navigational instrument based on a 60° arc (one-sixth of a circle) that is used to measure the height of a celestial object above the horizon.

sidereal year the year measured by the positions of the stars. *Compare* **tropical year.**

sloop a small sailboat with a single mast and fore-and-aft sails.

solstice Midwinter or Midsummer Day.

Spanish Main the coast of Spanish America from Florida all the way to the mouth of the Orinoco River.

starboard the right side of a ship when facing forward.

stem the curved piece of timber at the forward end of the ship that is an extension of the keel and to which the sides of the ship are attached; the opposite of **stern.**

stepped of a ship's mast, fitted into a socket attached to the hull.

stern the rearmost part of a ship; the opposite of **stem.**

strake one of the planks that are joined end-to-end to build the hull of a vessel.

tack to change direction in order to allow a headwind to blow from the side so a sailing ship can advance against it.

taffrail log (patent log) a ship's **log** consisting of a metal device, shaped like a torpedo with vanes at the rear end that were free to rotate as the log was towed through the water. It was lowered from the stern and the ship's speed was calculated from the number of turns made by the vanes in a given time.

tiller a pole inserted horizontally through the rudder shaft.

tramp steamer a merchant ship that does not sail to a regular schedule visiting particular ports, but obtains cargoes wherever it can.

trireme a **galley** with three rows of oars on each side.

tropical year the year measured as the interval between **equinoxes.** *Compare* **sidereal year.**

tundra vegetation typical of climates with a long, very cold winter and short summer with temperatures only a little above freezing. The vegetation comprises clump-forming grasses, sedges, and species of trees such as birch and willow that grow to only about 24 inches (60 cm) tall.

wind rose a diagram that shows the prevailing winds at a particular place. It consists of a circle with straight lines radiating from the center. Each line represents a wind direction and the length of each line indicates the frequency with which the wind blew from that direction over a reference period, usually of one month or one year.

windward side the side exposed to the wind.

yard the rigid support from which a sail hangs.

FURTHER RESOURCES

Allaby, Michael. *Deserts.* New York: Facts On File, revised ed. 2008. One of the Ecosystem set of books, *Deserts* includes accounts of the lives and achievements of several explorers of deserts and polar regions, and also describes desert caravans.

Brown, George Mackay. *Vinland.* London: John Murray, 1992. A novel by a major Scottish writer describing the life of Ranald Sigmundson, born in Orkney when the islanders were struggling between their Viking past and their Christian future.

Graham, Eric J. *Seawolves: Pirates and the Scots.* Edinburgh: Birlinn, 2005. A scholarly but highly readable account of the brief heyday of piracy.

Jardine, Lisa. *Worldly Goods: A New History of the Renaissance.* London: Macmillan, 1996. A description of the rise of mercantilism and its contribution to European prosperity by a leading historian.

Lamb, H. H. *Climate, History and the Modern World.* New York: Routledge, 2nd ed. 1995. An account of the way climate change has influenced history, including Viking voyages and settlements, by one of the leading historians of climate.

Lewis, David and Derek Oulton. *We, the Navigators: The Ancient Art of Landfinding in the Pacific.* Honolulu: University of Hawaii Press, 1994. A short book describing in detail the construction and use of the traditional boats used throughout the Pacific region.

Monmonier, Mark. *Rhumb Lines and Map Wars: A Social History of the Mercator Projection.* Chicago: University of Chicago Press, 2004. A lucid account of the development of Mercator's map projection and of its predecessors and rivals by the Distinguished Professor of Geography at Syracuse University.

Peiser, Benny. "From Genocide to Ecocide: The Rape of Rapa Nui." *Energy & Environment* (16) 3 & 4, 2005. Brentwood, Essex: Multi-Science Publishing Co. Ltd. A scholarly account of how the tale of the self-destruction of the Rapa Nui islanders' culture conceals the true story of its destruction by Europeans.

Polo, Marco. *The Travels,* translated by Ronald Latham. New York: Penguin Books, 1958. The text of Marco Polo's book with an introduction by the translator.

Souhami, Diana. *Selkirk's Island.* London: Weidenfeld & Nicolson, 2001. The true story of Alexander Selkirk, on whom Daniel Defoe based the fictional character of Robinson Crusoe.

WEB SITES

King Alfred the Great. "Jubilee Edition of the Whole Works, with Preliminary Essays Illustrative of the History, Arts, and Manners, of the Ninth Century." Oxford and Cambridge: J. F. Smith and Co. for the Alfred Committee, 1852. Available online. URL: http://www.archive.org/stream/wholeworkswithpr02alfruoft/wholeworkswithpr02alfruoft_dj vu.txt. Accessed May 22, 2009. The full text of this work, which includes Alfred's account of the voyages of Othar.

Asher, Michael. "Sir Wilfred Thesiger." *The Guardian,* August 27, 2003. Available online. URL: http://www.guardian.co.uk/travel/2003/aug/27/booksobituaries.obituaries. Accessed July 22, 2009. An obituary of Thesiger.

Cambridge and Boston Press. "Phoenicians: The Phoenician Experience in Phoenicia and Lebanon." Available online. URL: http://www.phoenician.org/. Accessed May 20, 2009. A large site that uses up-to-date archaeological information to tell the story of the Phoenicians.

Dollinger, André. "Ships and Boats: The archaeological evidence." Available online. URL: http://reshafim.org.il/ad/egypt/timelines/topics/navigation.htm. Updated January 2009. Accessed May 6, 2009. An account of the construction and performance of ancient Egyptian boats and seagoing ships.

Food and Agriculture Organization. "Bay of Bengal Programme: Small-Scale Fisherfolk Communities; Development of Outrigger Canoes in Sri Lanka." Available online. URL: ftp://ftp.fao.org/docrep/fao/007/ae440e/ae440e00.pdf. FAO (UN) report on the use of outrigger canoes.

Herodotus. "The History of Herodotus." Translated by George Rawlinson. Available online. URL: http://classics.mit.edu/Herodotus/history.html. Accessed June 15, 2009. The complete text of all nine books of Herodotus's work, written in 440 B.C.E.

Ifland, Peter. "The History of the Sextant." University of Coimbra, Portugal. Available online. URL: http://www.mat.uc.pt/~helios/Mestre/Novemb00/H61iflan.htm. Accessed June 12, 2009. The text of a lecture delivered on October 3, 2000, in which Dr. Ifland outlined the development of navigational instruments that culminated in the sextant.

NASA. "Earth's Inconstant Magnetic Field." Available online. URL: http://science.nasa.gov/headlines/Y2003/29dec_magneticfield.htm. Accessed June 5, 2009. A simple explanation of the Earth's magnetic field and the location of the magnetic poles.

———. "The Apollo Program." NASA History Division. Available online. URL: http://history.nasa.gov/apollo.html. Accessed July 23, 2009. A full history of the Apollo program.

National Maritime Museum. "Explorers and Leaders: Sir John Franklin (1786–1847)." Available online. URL: http://www.nmm.ac.uk/explore/sea-and-ships/facts/explorers-and-leaders/sir-john-franklin-(178 6–1847). Published February 1, 2000. Accessed July 6, 2009. An account of the fate of Franklin's expedition to find the Northwest Passage.

Odoric, Friar. "The Journal of Friar Odoric." From Hakluyt, Richard, *Principal Navigations, Voyages, Traffiques and Discoveries of the English Nation* (Goldsmid edition). eBooks@Adelaide. Available online. URL: http://ebooks.adelaide.edu.au/h/hakluyt/voyages/odoric/. Accessed June 18, 2009. The complete text of Odoric's travels, in late-16th century English.

Planetary Society, The. Home page available online. URL: http://www.planetary.org/home/. Accessed July 24, 2009. The Planetary Society is the world's largest public organization promoting space exploration.

Ross, Julie Megan. "A Paleoethnobotanical Investigation of Gerden Under Sandet, a Waterlogged Norse Farm Site, Western Settlement, Greenland." University of Alberta. Available online. URL: http://www.collectionscanada.gc.ca/obj/s4/f2/dsk2/ftp04/mq22551.pdf. Accessed May 25, 2009. A detailed account of the way of life of the Norse settlers in Greenland in the Middle Ages.

Ryan, Donald P. "The Ra Expeditions Revisited." Available online. URL: http://www.plu.edu/~ryandp/RAX.html. Accessed May 15, 2009. A summary of the reasoning behind the Kon-Tiki and Ra Expeditions.

SETI Institute. Available online. URL: http://www.seti.org/Page.aspx?pid=1345. Accessed July 27, 2009. The home page of the SETI (Search for Extraterrestrial Intelligence) Institute.

South-Pole.com. Available online. URL: http://www.south-pole.com/homepage.html. Accessed July 16, 2009. Biographies of many of the explorers of Antarctica.

Wild, Oliver. "The Silk Road." Available online. URL: http://www.ess.uci.edu/~oliver/silk.html. Accessed May 28, 2009. A detailed account of the history and route of the Silk Road.

Xenophon. "Anabasis, or *March Up Country*." Available online. URL: http://www.fordham.edu/halsall/ancient/xenophon-anabasis.html. Accessed June 16, 2009. An English translation of the full text of Xenophon's work.

INDEX

Note: *Italic* page numbers indicate illustrations or historical maps; page numbers followed by *m* indicate functional maps.

N

O

P

X

Y

Z